STEWARDSHIP

OF FEDERAL

FACILITIES

A Proactive Strategy for Managing the Nation's Public Assets

Committee to Assess Techniques for Developing
Maintenance and Repair Budgets for Federal Facilities

Board on Infrastructure and the Constructed Environment

Commission on Engineering and Technical Systems

National Research Council

NATIONAL ACADEMY PRESS
Washington, D.C. 1998

NATIONAL ACADEMY PRESS • 2101 Constitution Avenue, N.W. • Washington, DC 20418

NOTICE: The project that is the subject of this report was approved by the Governing Board of the National Research Council, whose members are drawn from the councils of the National Academy of Sciences, the National Academy of Engineering, and the Institute of Medicine. The members of the committee responsible for the report were chosen for their special competences and with regard for appropriate balance.

The National Academy of Sciences is a private, nonprofit, self-perpetuating society of distinguished scholars engaged in scientific and engineering research, dedicated to the furtherance of science and technology and to their use for the general welfare. Upon the authority of the charter granted to it by the Congress in 1863, the Academy has a mandate that requires it to advise the federal government on scientific and technical matters. Dr. Bruce Alberts is president of the National Academy of Sciences.

The National Academy of Engineering was established in 1964, under the charter of the National Academy of Sciences, as a parallel organization of outstanding engineers. It is autonomous in its administration and in the selection of its members, sharing with the National Academy of Sciences the responsibility for advising the federal government. The National Academy of Engineering also sponsors engineering programs aimed at meeting national needs, encourages education and research, and recognizes the superior achievements of engineers. Dr. William Wulf is president of the National Academy of Engineering.

The Institute of Medicine was established in 1970 by the National Academy of Sciences to secure the services of eminent members of appropriate professions in the examination of policy matters pertaining to the health of the public. The Institute acts under the responsibility given to the National Academy of Sciences by its congressional charter to be an adviser to the federal government and, upon its own initiative, to identify issues of medical care, research, and education. Dr. Kenneth I. Shine is President of the Institute of Medicine.

The National Research Council was organized by the National Academy of Sciences in 1916 to associate the broad community of science and technology with the Academy's purposes of furthering knowledge and advising the federal government. Functioning in accordance with general policies determined by the Academy, the Council has become the principal operating agency of both the National Academy of Sciences and the National Academy of Engineering in providing services to the government, the public, and the scientific and engineering communities. The Council is administered jointly by both Academies and the Institute of Medicine. Dr. Bruce Alberts and Dr. William Wulf are chairman and vice chairman, respectively, of the National Research Council.

This study was supported by Contract No. S-FBOAD-94-C-0023 between the National Academy of Sciences and the Department of State on behalf of the Federal Facilities Council. Any opinions, findings, conclusions, and recommendations expressed in this publication are those of the authors and do not necessarily reflect the view of the organizations or agencies that provided support for this project.

International Standard Book Number: 0-309-06189-X
Library of Congress Catalog Card Number: 98-87971

Available in limited supply from: Federal Facilities Council, HA 274, 2101 Constitution Avenue, N.W., Washington, D.C. 20418, (202) 334-3374

Additional copies available for sale from: National Academy Press, 2101 Constitution Avenue, NW, Box 285, Washington, D.C. 20055, 1-800-624-6242 or (202) 334-3313 (in the Washington metropolitan area). http://www.nap.edu

iv

Preface

Buildings and other constructed facilities represent investments made by owners in anticipation of the shelter and services they will provide to the people using them and the activities performed within them. Easily recognized facilities like the White House, the U.S. Capitol, and the Washington Monument are important symbols of the American government at home and abroad. Historic and architecturally significant facilities, however, represent only a small fraction of the more than 500,000 buildings and other structures, and their associated infrastructure, that have been acquired by the federal government to support defense and foreign policy missions; house historic, cultural, and artistic artifacts; serve as workplaces for scientists, engineers, educators, and researchers; and provide services to the American public.

Stakeholders in these federal facilities include the American public, whose tax dollars have been invested in acquiring and maintaining them and who regularly use and depend on the services supported by these facilities but who are increasingly critical of the cost of government; Congress, which appropriates the funding to acquire and maintain them; federal employees, who conduct the business of government; and foreign tourists, who visit these facilities.

The ownership of real property entails an investment in the present and a commitment to the future. Ownership of facilities by the federal government, or any other entity, represents an obligation that requires not only money to carry out that ownership responsibly, but also the vision, resolve, experience, and expertise to ensure that resources are allocated effectively to sustain that investment. Recognition and acceptance of this obligation is the essence of stewardship. Yet, despite the millions of stakeholders and the expenditure of hundreds of billions of taxpayer dollars in federal facilities, we have not been good stewards of our public buildings. The continuing deterioration of federal facilities is apparent to the most casual observer and has been documented by numerous studies. Still, little has been done to check the decline, and few people in the government are willing to accept responsibility for it.

Inadequate funding for maintenance and repair programs in the federal government is a long-standing problem. Plans and programs for maintenance have received little support from executive or legislative decision makers for several reasons. First, there is a tacit assumption that maintenance can always be put off for a month, a year, or even five years in favor of current operations and more visible projects with seemingly higher payoffs. Second, managers of federal agencies have generally failed to convince the public or political decision makers that funds invested in preventive and timely maintenance will be cost effective, will protect the quality and functionality of facilities, and will protect the taxpayers' investment. Thus, decision makers, who tend to have short-term outlooks, have not often been swayed to support actions with results that are realized over the long term.

Properly maintained federal facilities are not a luxury. They are critical to the effective performance of government agencies' missions and the provision of government services to the public. Inadequate maintenance in public buildings can have serious and costly consequences: damage caused by leaking roofs, burst pipes, and malfunctioning ventilation systems can cause disruptions of work, computer and other technological breakdowns, risks to occupants' health and safety, lost productivity, and millions of dollars in emergency repairs.

We cannot continue to ignore the consequences of undermaintaining our public buildings. Disinvestment is causing an inexorable and unacceptable degradation of the nation's public assets and a decline in the functionality and quality of federal facilities. The investment made in these assets warrants sustained, appropriately timed and targeted maintenance. Responsible investment in, and stewardship of, public buildings would optimize their service life, would be cost effective over the life of the facilities, and would contribute to a safer, healthier, more productive environment for the American public, foreign visitors, employees, and the officials who use these facilities every day.

In these times of decreasing budgets and downsizing, many interests must compete for limited resources. This study is not simply a call for increased expenditures for the maintenance and repair of federal facilities. It recommends a rationale and strategy for facility management, maintenance, and accountability for stewardship that will optimize limited resources while protecting the value and functionality of the nation's public buildings and other constructed facilities.

We have a significant opportunity to strategically redirect federal facilities management and maintenance practices for the twenty-first century. This will require long-term vision, commitment, leadership, and stewardship by both federal decision makers and agency managers. The results will be a significant improvement in the quality and performance of our federal facilities, lower overall maintenance costs, and protection of our public investment.

JACK E. BUFFINGTON
Chair, Committee to Assess Techniques for
Developing Maintenance and Repair Budgets
for Federal Facilities

Acknowledgments

This report has been reviewed by individuals chosen for their diverse perspectives and technical expertise, in accordance with procedures approved by the National Research Council's (NRC's) Report Review Committee. The purpose of this independent review is to provide candid and critical comments that will assist the authors and the NRC in making the published report as sound as possible and to ensure that the report meets institutional standards for objectivity, evidence, and responsiveness to the study charge. The contents of the review comments and draft manuscript remain confidential to protect the integrity of the deliberative process. We wish to thank the following individuals for their participation in the review of this report:

Mr. Harry Hatry, The Urban Institute, Washington, D.C.
Dr. Cameron Gordon, University of Southern California-Washington Center, Washington, D.C.
Dr. Harvey Kaiser, Consultant, Syracuse, New York
Mr. Douglas Kincaid, Applied Management Engineering, Virginia Beach, Virginia
Mr. Harold Odom, Florida Department of Management Services, Tallahassee
Dr. Alan Steiss, University of Michigan, Ann Arbor

Although the individuals listed above have provided many constructive comments and suggestions, responsibility for the final content of this report rests solely with the authoring committee and the NRC.

Contents

Executive Summary

Since its establishment in 1789, the federal government has constructed and acquired buildings, other facilities, and their associated infrastructures to support the conduct of public policy, defend the national interest, and provide services to the American public. Today, the federal facilities inventory comprises more than 500,000 buildings and structures, as well as the power plants, utility distribution systems, roads, and other infrastructure required to support them. Federal facilities are located in all 50 states, U.S. territories, and more than 160 foreign countries. They span decades, sometimes centuries, of building design and construction technologies, support a myriad of government functions, and represent the investment of more than 300 billion tax dollars.[1]

Federal facilities embody significant investments and resources and therefore constitute a portfolio of public assets. These buildings and structures project an image of American government at home and abroad, contribute to the architectural and socioeconomic fabric of their communities, and support the organizational and individual performance of federal employees conducting the business of government. These assets must be well maintained to operate adequately and cost effectively, to protect their functionality and quality, and to provide a safe, healthy, productive environment for the American public, elected officials, federal employees, and foreign visitors who use them every day.

Despite the historic, architectural, cultural, and functional importance of, and the economic investment in, federal facilities, studies by the General Accounting Office (GAO) and other federal government agencies indicate that the physical

---

[1]As of fiscal year 1996, federal agencies reported $215.5 billion of investment in structures and facilities and almost $82 billion of construction in progress (GAO, 1998).

1

condition of this portfolio of public assets is deteriorating. Many necessary repairs were not made when they would have been most cost effective and have become part of a backlog of deferred maintenance. In addition, a large proportion of federal facilities are more than 40 years old. As wear and tear on buildings and their systems increases, the need for maintenance and repair to sustain their performance and functionality also increases. Federal agency program managers, the GAO, and research organizations have all reported that the funding allocated for the repair and maintenance of federal facilities is insufficient.

Although there is no single, agreed upon guideline to determine how much money is, in fact, necessary to maintain public buildings, a 1990 report of the National Research Council, Committing to the Cost of Ownership: The Maintenance and Repair of Public Buildings, did recommend that, "An appropriate budget allocation for routine M&R [maintenance and repair] for a substantial inventory of facilities will typically be in the range of 2 to 4 percent of the aggregate current replacement value of those facilities" (NRC, 1990). This guideline has been widely quoted in the facilities management literature. During the course of the present study, federal agency representatives indicated that the funding they receive for maintenance and repair of their agencies' facilities is less than 2 percent of the aggregate current replacement value of their facilities inventories.

In an environment of inadequate and declining resources, federal facilities program managers face a number of challenges:

- extending the useful life of aging facilities
- altering or retrofitting facilities to consolidate space or accommodate new functions and technologies
- meeting evolving facility-related standards for safety, environmental quality, and accessibility
- maintaining or disposing of excess facilities created through military base closures and realignments, downsizing, or changing demographics
- finding innovative ways and technologies to maximize limited resources

To help federal agencies meet these challenges and optimize available resources, the sponsoring agencies of the Federal Facilities Council requested that the National Research Council review current federal practices for planning, budgeting, and implementing facility maintenance and repair programs and (1) develop a methodology and rationale federal facilities program managers can use for the systematic formulation and justification of facility maintenance and repair budgets; (2) investigate the role of technology in performing automated condition assessments; and (3) identify staff capabilities necessary to perform condition assessments and develop maintenance and repair budgets.

The study was conducted under the auspices of the Board on Infrastructure and the Constructed Environment by a committee of recognized experts in budgeting, facilities operations and maintenance, decision science, engineering economics, and building and facilities technology. Many of the committee members have professional experience with the management of large facilities portfolios. In addition to their own expertise, they were assisted by representatives of federal agencies, private sector organizations, and individuals who provided information on current budgeting, financial, maintenance, and building engineering practices in the federal government and the private sector.

Throughout this study, the committee was hampered by a lack of published data related to federal facilities inventories, programs, and practices. Accurate counts of basic items, such as the total number of federal facilities, the age of facilities, expenditures for maintenance and repair, were not available. (This issue is addressed in the study's findings and recommendations.) The committee also found that current maintenance and repair budgeting procedures, definitions, and accounting have advanced little since 1990. For information on the physical condition of federal facilities, maintenance and repair budgeting, condition assessment practices, deferred maintenance, and related topics, the committee relied heavily on reports of the GAO, briefings by federal agency program managers, and personal experience.

The committee began task 1 with the idea that it could develop a methodology for the systematic formulation of maintenance and repair budgets. However, the current state of practice, the general lack of data, and the lack of research results, in particular, precluded the development of a methodology per se. Instead, the committee identified methods, principles, and strategies that, if implemented, can be used to develop a methodology for the systematic formulation of maintenance and repair budgets in the future. In approaching task 2, the committee reviewed federal agency condition assessment practices and the role of technology in developing automated condition assessments. The committee found that existing sensor and microprocessor technologies have the potential to monitor and manage a range of building conditions and environmental parameters, but, for economic and other reasons, they have not been widely deployed. The committee also reviewed staff capabilities necessary to the performance of condition assessments and the development of maintenance and repair budgets (task 3). The committee found that both require adequate training for staff to foster effective decision making in facilities management, condition assessments, and maintenance and repair budgeting.

Federal government standards for internal oversight and control require that agencies safeguard all of the assets entrusted to them. This report seeks to foster accountability for the stewardship (i.e., responsible care) of federal facilities at all levels of government. The committee's findings and recommendations are presented below.

FINDINGS

Finding 1. Based on the information available to the committee, the physical condition of the federal facilities portfolio continues to deteriorate, and many federal buildings require major repairs to bring them up to acceptable quality, health, and safety standards.

Finding 2. The underfunding of facilities maintenance and repair programs is a persistent, long-standing problem. Federal operating and oversight agencies report that agencies have excess, aging facilities and insufficient funds to maintain, repair, or update them. Information provided to the committee indicated that agencies are receiving less than 2% of the aggregate current replacement value of their facilities inventories for maintenance and repair.

Finding 3. Federal government processes and practices are generally not structured to provide for effective accountability for the stewardship (i.e., responsible care) of federal facilities.

Finding 4. Buildings and facilities are durable assets that contribute to the effective provision of government services and the fulfillment of agency missions. However, the relationship of facilities to agency missions has not been recognized adequately in federal strategic planning and budgeting processes.

Finding 5. Maintenance and repair expenditures generally have less visible or less measurable benefits than other operating programs. Facilities program managers have found it difficult to make compelling arguments to justify these expenditures to public officials, senior agency managers, and budgeting staff.

Finding 6. Budgetary pressures on federal agency managers encourage them to divert potential maintenance and repair funds to support current operations, to meet new legislative requirements, or to pay for operating new facilities coming on line.

Finding 7. It is difficult, if not impossible, to determine how much money the federal government as a whole appropriates and spends for the maintenance and repair of federal facilities because definitions and calculations of facilities-related budget items, methodologies for developing budgets, and accounting and reporting systems for tracking maintenance and repair expenditures, vary.

Finding 8. There is evidence that some agencies own and are responsible for more facilities than they need to support their missions or than they can maintain with current or projected budgets.

Finding 9. Federal facilities program managers are being encouraged to be more businesslike and innovative, but current management, budgeting, and financial processes have disincentives and institutional barriers to cost-effective facilities management and maintenance practices.

Finding 10. Performance measures to determine the effectiveness of maintenance and repair expenditures have not been developed within the federal government. Thus, it is difficult to identify best practices for facilities maintenance and repair programs across or within federal agencies.

Finding 11. Based on the information available to the committee, federal condition assessment programs are labor intensive, time consuming, and expensive. Agencies have had limited success in making effective use of the data they gather for timely budget development or for the ongoing management of facilities.

Finding 12. Organizational downsizing has forced facilities program managers to look increasingly to technology solutions to provide facilities-related data for decision making and for performing condition assessments.

Finding 13. Existing sensor and microprocessor technologies have the potential to monitor and manage a range of building conditions and environmental parameters, but, for economic and other reasons, they have not been widely deployed.

Finding 14. Training for staff is a key component of effective decision making, condition assessments, and the development of maintenance and repair budgets.

Finding 15. Only a limited amount of research has been done on the deterioration/failure rates of building components or the nonquantitative implications of building maintenance (or lack thereof). This research is necessary to identify effective facilities management strategies for achieving cost savings, identifying cost avoidances, and providing safe, healthy, productive work environments.

Finding 16. Greater accountability for the stewardship of facilities is necessary at all levels of the federal government. Accountability includes being held responsible for the condition of facilities and for the allocation, tracking, and effective use of maintenance and repair funds.

The committee recommends that the government take the following actions (which are not in any particular order of priority).

RECOMMENDATIONS

Recommendation 1. The federal government should plan strategically for the maintenance and repair of its facilities in order to optimize available resources, maintain the functionality and quality of federal facilities, and protect the public's investment. A recommended strategic framework of methods, practices, and strategies for the proactive management and maintenance of the nation's public assets is summarized on Figure ES-1 (Findings 1 and 2).

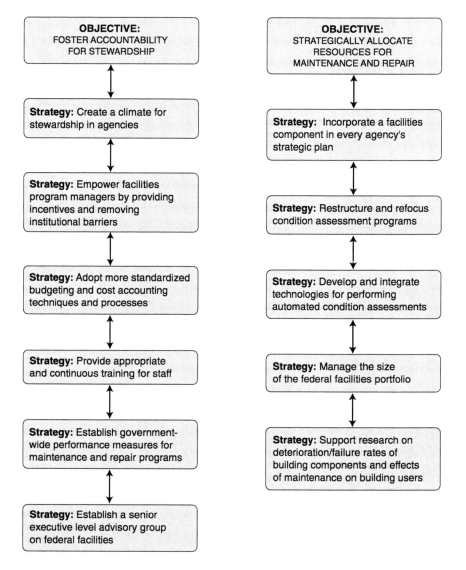

GOAL
Protect and Enhance the Functionality and Quality
of the Federal Facilities Portfolio

OBJECTIVE:
FOSTER ACCOUNTABILITY
FOR STEWARDSHIP

OBJECTIVE:
STRATEGICALLY ALLOCATE
RESOURCES FOR
MAINTENANCE AND REPAIR

Strategy: Create a climate for stewardship in agencies

Strategy: Incorporate a facilities component in every agency's strategic plan

Strategy: Empower facilities program managers by providing incentives and removing institutional barriers

Strategy: Restructure and refocus condition assessment programs

Strategy: Adopt more standardized budgeting and cost accounting techniques and processes

Strategy: Develop and integrate technologies for performing automated condition assessments

Strategy: Provide appropriate and continuous training for staff

Strategy: Manage the size of the federal facilities portfolio

Strategy: Establish government-wide performance measures for maintenance and repair programs

Strategy: Support research on deterioration/failure rates of building components and effects of maintenance on building users

Strategy: Establish a senior executive level advisory group on federal facilities

FIGURE ES-1   Strategic framework for the maintenance and repair of federal facilities.

Recommendation 2. The government should foster accountability for the stewardship of federal facilities at all levels. Facilities program managers at the agency level should identify and justify the resources necessary to maintain facilities effectively and should be held accountable for the use of these resources (Findings 1, 2, 3 and 16).

Recommendation 3. At the executive level, an advisory group of senior level federal managers, other public sector managers, and representatives of the nonprofit and private sectors should be established to develop policies and strategies to foster accountability for the stewardship of facilities and to allocate resources strategically for their maintenance and repair (Findings 1, 2, 3 and 16).

Recommendation 4. Facility investment and management should be directly linked to agency mission. Every agency's strategic plan should include a facilities component that links facilities to agency mission and establishes a basis and rationale for maintenance and repair budget requests (Finding 4).

Recommendation 5. The government should adopt more standardized budgeting and cost accounting techniques and processes to facilitate tracking of maintenance and repair funding requests, allocations, and expenditures and reflect the total costs of facilities ownership. The committee developed an illustrative template as shown in Figure ES-2 (Findings 3, 5, 6, 7 and 16).

Recommendation 6. Government-wide performance measures should be established to evaluate the effectiveness of facilities maintenance and repair programs and expenditures (Finding 10).

Recommendation 7. Facilities program managers should be empowered to operate in a more businesslike manner by removing institutional barriers and providing incentives for improving cost-effective use of maintenance and repair funds. The carryover of unobligated funds and the establishment of revolving funds for nonrecurring maintenance needs should be allowed if they are justified (Findings 3 and 9).

Recommendation 8. Long-term requirements for maintenance and repair expenditures should be managed by reducing the size of the federal facilities portfolio. New construction should be limited, existing buildings should be adapted to new uses, and the ownership of unneeded buildings should be transferred to other public or private organizations. Facilities that are functionally obsolete, are not needed to support an agency's mission, are not historically significant, and are not suitable for transfer or adaptive reuse should be demolished whenever it is cost effective to do so (Findings 2, 8 and 16).

Facilities Management- Related Activities	Included in 2-4% Benchmark	Funding Category and Comments
A. **Routine Maintenance, Repairs, and Replacements** • recurring, annual maintenance and repairs including maintenance of structures and utility systems, (including repairs under a given $ limit, e.g., $150,000 to $500,000 exclusive of furniture and office equipment) • roofing, chiller/boiler replacement, electrical/lighting, etc. • preventive maintenance • preservation/cyclical maintenance • deferred maintenance backlog • service calls	Yes	Annual operating budget
B. **Facilities-Related Operations** • custodial work (i.e., services and cleaning) • utilities (electric, gas, etc./plant operations) • snow removal • waste collection and removal • pest control • security services • grounds care • parking • fire protection services	No	Annual operating budget
C. **Alterations and Capital Improvements** • major alterations to subsystems, (e.g., enclosure, interior, mechanical, electrical expansion) that change the capacity or extend the service life of a facility • minor alterations (individual project limit to be determined by agency $50,000 to $1 million)	No	Various funding sources, including no year, project- based allocations such as revolving funds, carryover of unobligated funds, fund- ing resulting from cost savings or cost avoidance strategies
D. **Legislatively Mandated Activities** • improvements for accessibility, hazardous materials removal, etc.	No	Various sources of funding
E. **New Construction and Total Renovation Activities**	No	Project-based allocations separate from operations and maintenance budget. Should include a life-cycle cost analysis prior to funding
F. **Demolition Activities**	No	Various sources of funding

FIGURE ES-2   Illustrative template to reflect the total costs of facilities ownership.

Recommendation 9. Condition assessment programs should be restructured to focus first on facilities that are critical to an agency's mission; on life, health, and safety issues; and on building systems that are critical to a facility's performance. This will optimize available resources, provide timely and accurate data for formulating maintenance and repair budgets, and provide critical information for the ongoing management of facilities (Findings 4 and 11).

Recommendation 10. The government should provide appropriate and continuous training for staff that perform condition assessments and develop and review maintenance and repair budgets to foster informed decision making on issues related to the stewardship of federal facilities and the total costs of facilities ownership (Findings 14 and 16).

Recommendation 11. The government and private industry should work together to develop and integrate technologies for performing automated facility condition assessments and to eliminate barriers to their deployment (Findings 11, 12 and 13).

Recommendation 12. The government should support research on the deterioration/failure rates of building components and the nonquantitative effects of building maintenance (or lack thereof) in order to develop quantitative data that can be used for planning and implementing cost-effective maintenance and repair programs and strategies and for better understanding the programmatic effects of maintenance on mission delivery and on building users' health, safety, and productivity (Findings 12 and 15).

REFERENCES

GAO (General Accounting Office). 1998. Deferred Maintenance Reporting: Challenges to Implementation. Report to the Chairman, Committee on Appropriations, U.S. Senate. AIMD-98-42. Washington, D.C.: Government Printing Office.
NRC (National Research Council). 1990. Committing to the Cost of Ownership: Maintenance and Repair of Public Buildings. Building Research Board, National Research Council. Washington, D.C.: National Academy Press.

1

Introduction

Buildings and other constructed facilities are investments made by owners in anticipation of the services they will provide and the activities they will support. To serve specific functions and missions and generally conduct its business, the federal government has built or acquired more than 500,000 buildings, facilities, and their associated infrastructures worldwide (i.e., roads, utility plants, distribution systems, and the like). Government facilities are used to defend the national interest; conduct foreign policy; house historic, cultural, and educational artifacts; pursue research; and provide services to the American public. Buildings of fundamental architectural or historical significance, such as the White House, the United States Capitol, and monuments to national heroes and events, symbolize the American government and heritage. Military installations, which are often the size of small cities, support the defense and protection of American interests at home and abroad. Embassy compounds house and provide workplaces for government employees conducting foreign policy and serving American citizens overseas. Archives, libraries, and museums are repositories for priceless and irreplaceable documents, literature, art, and artifacts that embody human culture and history. Research laboratories and space centers provide workplaces for scientists, engineers, and medical experts developing technologies, techniques, and medicines to improve the quality of life for current and future generations. Courthouses, prisons, hospitals, and administrative offices support the provision of a wide range of services to local communities. National park facilities provide recreational opportunities for citizens and foreign visitors.

Federal facilities comprise a portfolio of significant, durable public assets that reflect the investment of more than 300 billion tax dollars (Table 1-1).

TABLE 1-1 Reported Property, Plant, and Equipment (PP&E) by Federal Agencies for Fiscal Year 1996

Agency	Total PP&E ($ billions)	Percent of Total PP&E	Investment in Structures ($ billions)	Construction in Progress ($ billions)
U.S. Department of Defense	$773	80.5	$123.0	$63.5
Tennessee Valley Authority	30	3.2	22.2	.8
National Aeronautics and Space Administration	26	2.7	5.9	5.0
U.S. Department of Transportation	24	2.5	10.4	3.4
U.S. Department of Energy	22	2.3	11.9	3.7
U.S. Postal Service	18	1.9	8.3	1.6
U.S. Department of the Interior	17	1.7	15.9	—
General Services Administration	12	1.3	6.9	2.4
U.S. Department of Veterans Affairs	11	1.2	7.1	1.2
U.S. Department of Agriculture	9	.9	1.8	—
U.S. Department of State	5	.5	1.9	.3
All other agencies	14	1.4	—	—
Total*	$960	100.0	$215.5	$81.9

*Figures do not add precisely due to rounding off.

Source: Data from audited financial statements for 11 agencies. For all other agencies, data reported to the Department of the Treasury. This information was not independently verified by GAO. Source: GAO, 1998.

The investment in facilities supports an even larger investment in human resources. Industry and government studies have shown that the salaries paid to the occupants of a commercial or institutional building each year are of the same order of magnitude as the total costs of designing and constructing the building. Therefore, an "Improvement of the productivity of the occupants . . . is the most important performance characteristic for most constructed facilities" (NSTC, 1995).

LIFE CYCLES OF BUILDINGS

Buildings and other constructed facilities pass through a number of stages during their lifetimes: planning, design, construction, commissioning/occupancy, operation and use, renewal/revitalization, and disposal. Most constructed facilities are designed to provide at least a minimum acceptable level of shelter and service for 30 years. With proper management and maintenance[1] buildings may perform adequately for 40 to 100 years or more and may serve several different functions.

Buildings are complex structures with a number of separate but interrelated components. The components of the building "envelope" include roofs, walls, windows/doors, cladding materials (e.g., brick, stone, clapboard), and foundations. Critical servicing components include mechanical, electrical, plumbing, heating, air conditioning, ventilation, communications, fire, and safety systems. Each component must perform well to optimize a building's performance and service life and to provide a safe, healthy, and productive environment.

The service life, or period of time over which a building, component, or subsystem actually provides adequate performance, depends on many factors. The quality of a building's design, the durability of construction materials and component systems, the incorporated technology, the location and climate, the use and intensity of use, and damage caused by heavy storms, natural disasters, or human error all influence how well and how quickly a building ages and the amount of maintenance and repair a building requires over its life cycle. Although a building's performance inevitably declines because of aging, wear and tear, and functional changes, its service life can be optimized through adequate and timely maintenance and repairs, as illustrated in Figure 1-1. Conversely, when maintenance and repair activities are continuously deferred, the result can be an irreversible loss of service life.

---

[1]For this study, "maintenance" is defined as the upkeep of property and equipment, i.e., work necessary to realize the originally anticipated useful life of a fixed asset. "Repair" is defined as work to restore damaged or worn-out property to a normal operating condition. An effective maintenance and repair program includes several different types of activities that address different aspects/components and have different objectives. Activities include preventive maintenance, programmed major maintenance, predictive testing and inspection, routine repairs, and emergency service calls.

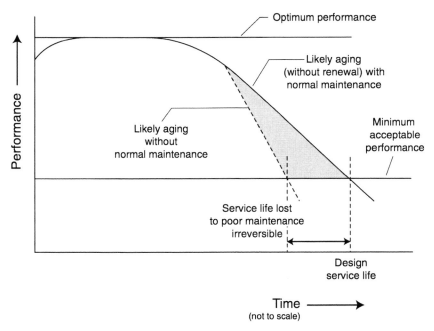

FIGURE 1-1 Effect of adequate and timely maintenance and repairs on the service life of a building. Source: NRC, 1993.

The total cost of ownership of a facility is the "total of all expenditures an owner will make over the course of the building's service lifetime" (NRC, 1990). Thus, an owner is responsible for funding not only planning, design, and construction, but also maintenance, repairs, replacements, alterations, and normal operations, such as heating, cooling, and lighting, and finally, demolition. Failure to recognize these costs and provide adequate maintenance and repair results in a shorter service life, more rapid deterioration, and higher operating costs over the life cycle of a building.

FACTORS CONTRIBUTING TO THE DETERIORATING CONDITION OF FEDERAL FACILITIES

Despite the historic, cultural, and architectural importance of, and economic investment in, federal facilities, evidence is mounting that the physical condition, functionality, and quality of the federal facilities portfolio is deteriorating. In response to Congressional inquiries, the General Accounting Office (GAO) has published a number of reports documenting the deterioration of federal facilities since 1990. These include NASA Maintenance: Stronger Commitment Needed to Curb Facility Deterioration (GAO, 1990), Federal Buildings: Actions Needed to Prevent Further Deterioration and Obsolescence (GAO, 1991), Federal Research: Aging Federal Laboratories Need Repairs and Upgrades (GAO, 1993), and National Parks: Difficult Choices Need to be Made About the Future of the Parks (GAO, 1995b). To cite only two examples from these reports, "at Ellis Island in New York, the nation's only museum devoted exclusively to immigration, 32 of 36 historic buildings have seriously deteriorated, and, according to park officials, about two-thirds of these buildings could be lost within 5 years if not stabilized." In one building used for storing cultural artifacts, "much of the collection is covered with dirt and debris from crumbling walls and peeling paint, and leaky roofs have caused water damage to many artifacts" (GAO, 1995a). A number of factors that contribute to the deteriorating condition of federal facilities, are described below.

Focus on First Costs

The deteriorating condition of federal facilities is attributable, in part, to the federal government's failure to recognize the total costs of facilities ownership. Although the "costs to operate and maintain a facility vary between 60 and 85 percent of its total ownership cost" (Christian and Pandeya, 1997), government budgeting practices have focused on the design and construction costs, or 5 to 10 percent of the total costs of ownership, the so-called "first" costs. (The remaining 5 to 35 percent of the costs of ownership include land acquisition, planning, renewal/revitalization, and disposal.)

The full life cycle costs of new facilities are not considered in the current federal budget process. Instead, only the projected design and construction costs appear as a separate line item for congressional consideration. The costs of operating and maintaining the new facility are not considered separately but become part of the agency's total operations and maintenance budget request, which includes funding for all existing facilities. The costs of designing and constructing a new facility, then, may receive considerable scrutiny during budget hearings, but the budget process is so structured that the 60 to 85 percent of the total costs, the costs of operating and maintaining the facility, do not receive the same scrutiny. Thus, the federal budget process is not structured to consider the total costs of facilities ownership.

Inadequate Funding for Maintenance and Repair

Inadequate funding for the maintenance and repair of public buildings at all levels of government and academia is a long-standing and well documented problem. A report by the National Research Council in 1990, Committing to the Cost of Ownership: Maintenance and Repair of Public Buildings, found that "Underfunding is a widespread and persistent problem that undermines maintenance and repair of public buildings" (NRC, 1990). A 1996 study by the Civil Engineering

Research Foundation reconfirmed this finding, noting that "underfunding of facilities maintenance and repair projects appears to be a widespread problem in both the public and private sectors" (CERF, 1996). On the subject of federal facilities, GAO has reported that, "mounting evidence shows that the federal government must also face up to the long-term consequences of inadequate capital investment in existing federal buildings" (GAO, 1991). More recently, GAO has found that "despite reductions in DoD's [U.S. Department of Defense] basing infrastructure, various DoD and service officials have continued to indicate that they still have excess, aging facilities and insufficient funding to maintain, repair, and update them" (GAO, 1997).

There is no single, agreed-upon guideline to determine how much money is adequate to maintain public buildings effectively. However, Committing to the Cost of Ownership: Maintenance and Repair of Public Buildings did recommend that, "An appropriate budget allocation for routine M&R [maintenance and repair] for a substantial inventory of facilities will typically be in the range of 2 to 4 percent of the aggregate current replacement value of those facilities" (NRC, 1990). This guideline has been widely quoted in the facilities management literature. During the course of this study, federal agency representatives who briefed the committee or completed questionnaires indicated that the funding they received annually for maintenance and repair was less than 2 percent of the aggregate current replacement value of their agencies' facilities inventories.[2] The National Aeronautics and Space Administration (NASA), for example, reported the maintenance and repair funding it currently receives to be about 1.3 percent of the current replacement value of all its facilities, and the Architect of the Capitol's Office reported funding of about 1.7 percent.

Deferred Maintenance

If funds are not available to address identified maintenance and repair needs, these projects may be deferred or delayed indefinitely. Deferred maintenance is defined in the Statement of Federal Financial Accounting Standards Number 6, Accounting for Property, Plant and Equipment, as "maintenance that was not performed when it should have been or was scheduled to be, and which, therefore, is put off or delayed for a future period" (GAO, 1998). Deferred maintenance, also called unfunded maintenance, backlog of maintenance and repair, or unaccomplished maintenance, is generally quantified as the estimated cost of the maintenance and repair needed to bring a facility up to a minimum acceptable condition. The significance of the existence of deferred maintenance is that it "implies that the quality and/or reliability of service provided by infrastructure on

---

[2]Agencies responding to the questionnaire included the U.S. Department of Energy, the Department of the Army/Installations, the International Broadcasting Bureau, the National Institute of Standards and Technology, the National Aeronautics and Space Administration, and the Office of the Air Force Civil Engineer.

which maintenance has been deferred is lower than it should be, and thus the infrastructure is not or will not later be adequately serving the public" (Urban Institute, 1994). A report by the American Public Works Association, Plan. Predict. Prevent. How to Reinvest in Public Buildings, found that "in the short-term, deferring maintenance will diminish the quality of building services. In the long-term, deferred maintenance can lead to shortened building life and reduced asset value" (APWA, 1992). In a series of reports, the GAO came to the following conclusions about the deferred maintenance of federal facilities:

> The Pentagon is a classic example of the federal government's failure to invest adequately in federal buildings . . . Needed structural repairs and upgrades to the Pentagon were deferred for more than a decade, and the General Services Administration (GSA) now estimates that its renovation will cost more than $1 billion and take at least 13 years to complete (GAO, 1991).

> Other federal buildings have been neglected . . . and now need major repairs and alterations to bring them up to acceptable quality, health and safety standards. The total number of federal buildings with deferred major repair and alteration requirements is unknown but our work suggests that the number may be substantial. Continuing to defer needed repairs and alterations accelerates deterioration and obsolescence and results in higher eventual costs to the government . . . (GAO, 1991).

> Most federal research laboratories are experiencing common problems with aging facilities—leaking roofs and gutters, drafty window frames, power outages, and poor ventilating systems that do not meet industry standards for air circulation . . . the eight agencies GAO reviewed reported backlogs of more than $3.8 billion in needed laboratory repairs (GAO, 1993).

> The overall level of visitor services offered by the National Park Service is deteriorating. Visitor services are being cut back and the condition of many trails, campgrounds, exhibits, and other facilities is declining. The Park Service estimates that since 1988, the backlog of deferred maintenance has more than doubled to $4 billion (GAO, 1995b).

The magnitude of the numbers cited by agencies indicates that significant needed maintenance and repairs have been deferred because of underfunding or other factors. Historically, public officials have not often found the arguments for maintenance and repair funding compelling and have called into question the methodologies used to define building deficiencies and to calculate the costs involved in repairing them.[3] One reason for this skepticism is that although "the amount of deferred maintenance is important in itself, without also including

---

[3]Fiscal year 1998 is the first year in which federal agencies are required to report periodically on deferred maintenance by disclosing deferred maintenance in agency financial statements. Previously, some but not all federal agencies kept inventories of building deficiencies and the funding required to eliminate them; others provided maintenance needs estimates for budgetary purposes and ad hoc reports.

information on the implications of deferral, public officials and the public will have considerable difficulty in interpreting the deferred maintenance figures" (Urban Institute, 1994). A second reason relates to the lack of a standard methodology for defining and quantifying deferred maintenance. The concern has been that inappropriate items have been included in the maintenance backlog to increase the overall estimate and argue for larger budget appropriations.

Agencies have also used different formulas or standards to compute the costs of eliminating the backlog. This situation may not be improved significantly by new reporting requirements of Federal Financial Accounting Standard Number 6 because under this standard "it is management's responsibility to . . . establish methods to estimate and report any material amounts of deferred maintenance" (GAO, 1998).

Aging of Facilities

The federal facilities portfolio includes structures that span centuries of different planning, design, construction, maintenance, management, and mission requirements. The average age of the federal facilities portfolio by square footage or by current replacement value is not known because accurate data are not available. However, it is safe to say that a large proportion of the facilities in the federal portfolio are already 40 to 50 years old. More than half of the 8,000 office buildings managed by the General Services Administration are more than 50 years old, and the U.S. State Department estimates the average age of facilities to be 39 years. Even in a "space age" agency like NASA, the average age of its facilities is approximately 40 years. As facilities age, wear and tear on building components increases, and electrical, mechanical, and other systems, begin to break down. The rate and onset of breakdowns increases if maintenance has been implemented haphazardly or not at all, and the operating condition deteriorates. Aging facilities require more, not less, maintenance and repair to keep them operating effectively.

Lack of Information to Justify Maintenance and Repair Budgets

In the federal budget and operations environment, facilities maintenance and repair is often deemed to be a low priority issue because facilities program managers do not have the information they need to present their case for funding to senior managers and public officials. "Interviews indicate that public officials, such as elected officials and chief administrative officers, find the most convincing and compelling information to be the future costs that can be avoided by undertaking early, preventive, or corrective maintenance activities" (Urban Institute, 1994). However, there is "very little study of the costs and implications of deferring maintenance . . . and cost avoidance information is lacking" (Urban Institute, 1994). Estimates of the implications of deferred maintenance on cost and quality of service are also lacking even though public officials "appear to believe such information to be of considerable use" (Urban Institute, 1994). Because information on maintenance and repair issues most convincing to public officials, particularly avoiding future costs, is not available, and because the information that is available, such as the backlog of deferred maintenance, is not compelling, facilities program managers have found it difficult to justify their maintenance and repair budget requests to senior executives and public officials.

Lack of Accountability for Stewardship

Buildings are durable assets constructed to last at least 30 years; but they are composed of a number of components with service lives of less than 10 years. Buildings themselves seldom fail in an obvious, catastrophic sense. The deterioration of individual components generally occurs over time and may not be readily apparent: detecting the incipient deterioration of roofs, mechanical and electrical systems, pipes, and foundations requires regular inspections by trained personnel. Once detected through regular inspections or condition assessments, relatively small problems can be repaired before they develop into much more serious problems through an adequately planned and funded maintenance program.

Because facility deterioration occurs over a long period of time, it may appear to senior executives and public officials that the maintenance and repair of facilities can always be deferred one more year without serious consequences in favor of more urgent operations that have greater visibility. Unless a roof actually falls in, senior managers are not likely to be held accountable for the condition of a facility in any given year. Yet they are held accountable for current operations. Consequently, public officials and senior executives have few incentives to practice effective stewardship of the federal facilities portfolio and are subject to few penalties if they do not.

CONSEQUENCES AND COSTS OF INADEQUATE MAINTENANCE

Continuously deferring adequate maintenance and repair can result in major damage to facilities, disruptions in service and business, and costly and serious health and safety consequences, as the following examples illustrate:

On May 26, 1989, at NASA's Lewis Research Center, a high pressure steam shutoff valve ruptured in the basement of the Library Services Building. The valve's failure was partially attributed to badly deteriorated piping supports in a steam line tunnel. Although the tunnel had inspection access holes, the piping supports were not included in a maintenance program. Heavy rains that flooded the tunnel caused steam to condense in the pipes and created a water hammer effect (a concussion of moving water against the sides of a containing pipe or vessel such as a steam pipe). The vibration of the poorly supported steam pipes caused the valve to rupture. In addition to damage to the valve and piping, high-pressure steam damaged two interior walls, an office, ceiling tiles, painted surfaces, and wall paneling throughout the building. The building was without steam service for 5 months, and the cost of repairs exceeded $1 million (GAO, 1990).

In August 1990, a small fire broke out at the Pentagon.

While the fire was being extinguished, an old, deteriorated 10-inch water pipe broke and flooded 350,000 square feet in the basement heating plant, the primary electrical switching room, and the Air Force's Communications Center. The basement heating plant was out of service for 2 days. Besides disrupting electrical power and interfering with Air Force operations, the flood resulted in approximately $500,000 in property damages (GAO, 1991).

The Nassif Building in Washington, D.C., a leased office facility of approximately 1.1 million square feet constructed in 1969, is the headquarters of the U.S. Department of Transportation. In October 1995, several employees complained of symptoms generally suggestive of "Sick Building Syndrome." As a result of these and many other complaints, DOT began an extensive investigation of the causes, including sampling and testing of the indoor environment, inspection of the building and mechanical systems, and medical examinations of affected employees. Significant findings from studies conducted in support of this investigation found that more than 50 percent of the roof drains on the building were leaking water into the tenth floor ceiling (DOT, 1996a), two of the four main ventilating units were so worn they were nonoperational and probably had been for some time, and that the overall quality of maintenance of the ventilation system was poor (DOT, 1996b). In a report on the indoor air quality of the building, the Occupational Safety and Health Administration recommended that maintenance plans and procedures be developed to respond to water leaks and the consequences of leaks and to inhibit microbial growth in the domestic hot water system (OSHA, 1995). A massive building cleaning and repair program required the temporary relocation offsite of personnel on a floor-by-floor basis. Not counting adverse health effects, losses in productivity, or any future legal claims, the cost to the government will exceed $13 million (Spillenkothen, 1997).

These incidents are not isolated instances of the consequences and costs of inadequate maintenance. They illustrate the conditions in many federal facilities and other public buildings. In all likelihood, incidents like these will happen more frequently in the future. A 1997 study of U.S. Department of Defense (DoD) facilities stated that officials at Army headquarters reported that "many of its [the Army's] installations are in a 'breakdown maintenance mode,' resulting in increases in emergency repairs and equipment breakdowns." At one installation, "emergency work orders increased from less than 300 for fiscal year 1992 to over 20,000 for fiscal year 1996," and "over 45 waterlines broke in fiscal year 1996." Navy headquarters officials reported that "funding levels allow only preventive maintenance on mission-critical systems, such as electrical and water pump distribution systems. The preventive maintenance is limited to inexpensive repairs that take as little as 15 minutes" (GAO, 1997).

BASIS FOR THIS STUDY

In the NRC report, Committing to the Cost of Ownership: Maintenance and Repair of Public Buildings, guidelines were recommended for developing maintenance and repair budgets for facilities in the absence of detailed cost estimates:

> M&R [maintenance and repair] budgets should be structured to identify explicitly the expenditures associated with routine M&R requirements and activities to reduce the backlog of deferred deficiencies. An appropriate budget allocation for routine M&R for a substantial inventory of facilities will typically be in the range of 2 to 4 percent of the aggregate current replacement value of those facilities (NRC, 1990).

Identified factors that can have a major influence on the appropriate level of M&R expenditures included building size and complexity, types of finishes, current age and condition, mechanical and electrical system technologies, telecommunications and security technologies, historic or community value, type of occupants or users, climatic severity, churn (i.e., tenancy turnover rates), criticality of role or function, ownership time horizon, labor prices, energy prices, materials prices, and distances between buildings in inventory. That report also suggested two additional areas of study: formalized condition assessment programs (including the role of technology); and staff capabilities to carry out condition assessment and M&R budgeting functions.

Based on the information available to the committee, no federal agency has consistently achieved a funding level equivalent of 2 to 4 percent of the aggregate current replacement value of its facilities inventory. In fact, other trends in the federal government have increased the pressure on maintenance and repair budgets. In an operating environment of declining resources, federal facilities program managers are faced with a number of challenges:

- maintaining a relatively stable number of facilities
- extending the useful life of aging facilities
- meeting evolving requirements for safety, environmental quality, and accessibility
- altering or retrofitting facilities to consolidate space or accommodate new functions or technologies
- overcoming institutional barriers to becoming more businesslike in their operations
- finding new ways to optimize available resources

Against this background, the sponsoring agencies of the Federal Facilities Council[4] determined that it would be appropriate to revisit the issue of budgeting techniques and activities for facility maintenance and repair and requested that a follow-up study to the 1990 report be done.

---

[4]The agencies that provided funding for this study through the Federal Facilities Council include the Office of the Air Force Civil Engineer, the Air National Guard, the U.S. Army Corps of Engineers,

STATEMENT OF TASK

The objectives of the follow-up study were to: (1) develop a methodology and rationale federal facilities program managers could use for the systematic formulation and justification of facility maintenance and repair budgets; (2) investigate the role of technology in performing automated condition assessments; and (3) identify staff capabilities necessary to perform condition assessments and develop maintenance and repair budgets. The Committee to Assess Techniques for Developing Maintenance and Repair Budgets for Federal Facilities was appointed by the National Research Council under the auspices of the Board on Infrastructure and the Constructed Environment. The committee members have a broad base of expertise including: government facilities budgeting and management, facilities operations and maintenance, public finance, building performance, facility technology and value engineering, computer applications for facility management, and condition assessments. The committee members have worked in federal, state, and local government agencies, private industry, and academia. (See Appendix A for biographical sketches.)

Throughout this study, the committee was hampered by a lack of published data related to federal facilities inventories, programs, and practices. Accurate counts of basic items such as the total number of federal facilities, the age of facilities, expenditures for maintenance and repair, were simply not available (see findings and recommendations). The committee also found that the state of practice in maintenance and repair budgeting procedures, definitions, and accounting had advanced little since 1990. For information on the physical condition of federal facilities, maintenance and repair budgeting, condition assessment practices, deferred maintenance, and related topics, the committee relied heavily on GAO reports, briefings by federal agency program managers, and personal experience.

The committee began task 1 with the idea that it could develop a methodology for the systematic formulation of maintenance and repair budgets. However, the current state of practice, the general lack of data, and the lack of research results in particular precluded the development of a methodology per se. The committee instead identified potential methods, principles, and strategies that, if implemented, could become the basis for the development of a methodology in the future.

In approaching task 2, the committee reviewed federal agency condition assessment practices and the role of technology in developing automated condition assessments. The committee found that existing sensor and microprocessor

---

the U.S. Department of Energy, the Naval Facilities Engineering Command, the Department of Veterans Affairs, the Food and Drug Administration, the General Services Administration, the National Aeronautics and Space Administration, the National Institutes of Health, the National Institute of Standards and Technology, the National Endowment for the Arts, the National Science Foundation, the Smithsonian Institution, the International Broadcasting Bureau, the U.S. Public Health Service, and the U.S. Postal Service.

technologies have the potential to monitor and manage a range of building conditions and environmental parameters, but, for economic and other reasons, they have not been widely deployed. In its review of the staff capabilities necessary to perform condition assessments and develop maintenance and repair budgets (task 3), the committee found that adequate training for staff is a key component in effective decision making in both facilities management and maintenance and repair budgeting.

ORGANIZATION OF THE REPORT

The succeeding chapters of this report address the statement of task in the following manner. Chapter 2 focuses on a wide range of issues related to the management and maintenance of federal facilities, including the federal budget process, the federal facilities portfolio, and the availability of maintenance and repair related data. Chapter 3 describes condition assessment practices, technologies, and issues. Chapter 4 presents a strategic framework for the maintenance and repair of federal facilities. Chapter 5 summarizes the study's findings and recommendations.

REFERENCES

APWA (American Public Works Association). 1992. Plan. Predict. Prevent. How to Reinvest in Public Buildings. Special Report #62. Chicago: American Public Works Association.

Christian, J., and A. Pandeya, 1997. Cost predictions of facilities. Journal of Management in Engineering 13(1): 52–61.

CERF (Civil Engineering Research Foundation). 1996. Level of Investment Study: U.S. Air Force Facilities and Infrastructure Maintenance and Repair. Washington, D.C.: Civil Engineering Research Foundation.

DOT (U.S. Department of Transportation). 1996a. Mechanical System Baseline Testing, U.S. Department of Transportation, Nassif Building, Washington, D.C. Chapters 8 and 9. Report by Summer Consultants, McLean, Virginia.

DOT. 1996b. Mechanical System Baseline Testing, U.S. Department of Transportation, Nassif Building, Washington, D.C., 31 March 1996. Report by Summer Consultants, McLean, Virginia.

GAO (General Accounting Office). 1990. NASA Maintenance: Stronger Commitment Needed to Curb Facility Deterioration. Report to the Chair, Subcommittee on VA, HUD and Independent Agencies, Committee on Appropriations, U.S. Senate. NSIAD-91-34. Washington, D.C.: Government Printing Office.

GAO. 1991. Federal Buildings: Actions Needed to Prevent Further Deterioration and Obsolescence. Report to the Chairman, Subcommittee on Public Works and Transportation, U.S. House of Representatives. GGD-91-57. Washington, D.C.: Government Printing Office.

GAO. 1993. Federal Research: Aging Federal Laboratories Need Repairs and Upgrades. Testimony. T-RCED-93-71. Washington, D.C.: Government Printing Office.

GAO. 1995a. National Parks: Difficult Choices Need to be Made About the Future of the Parks. Chapter Report. RCED-95-238. Washington, D.C.: Government Printing Office.

GAO. 1995b. National Parks: Difficult Choices Need to be Made About the Future of the Parks. Testimony. T-RCED-95-124. Washington, D.C.: Government Printing Office.

GAO. 1997. Defense Infrastructure: Demolition of Unneeded Buildings Can Help Avoid Operating Costs. Report to the Chair, Subcommittee on Military Installations and Facilities, Committee on National Security, U.S. House of Representatives. NSIAD-97-125. Washington, D.C.: Government Printing Office.

GAO. 1998. Deferred Maintenance Reporting: Challenges to Implementation. Report to the Chairman, Committee on Appropriations, U.S. Senate. AIMD-98-42. Washington, D.C.: Government Printing Office.

NRC (National Research Council). 1990. Committing to the Cost of Ownership: Maintenance and Repair of Public Buildings. Building Research Board, National Research Council. Washington, D.C.: National Academy Press.

NRC. 1993. The Fourth Dimension in Building: Strategies for Minimizing Obsolescence. Building Research Board, National Research Council. Washington, D.C.: National Academy Press.

NSTC (National Science and Technology Council). 1995. Construction and Building: Federal Research and Development in Support of the U.S. Construction Industry. Washington, D.C.: Government Printing Office.

OSHA (Occupational Safety and Health Administration). 1995. Indoor Air Quality Investigation, Nassif Building in Washington, D.C., Salt Lake City Technical Center Report.

Spillenkothen, M. 1997. Personal communication between Melissa Spillenkothen, Assistant Secretary for Administration, Department of Transportation, Washington, D.C. and Richard Little, Director, Board on Infrastructure and the Constructed Environment, National Research Council, Washington, D.C., May 7, 1997.

Urban Institute. 1994. Issues in Deferred Maintenance. Washington, D.C.: U.S. Army Corps of Engineers.

2

Related Issues

The issues related to the maintenance and repair of federal facilities are complex and include many interrelated components. This chapter focuses on three broad issues: the federal budget process; the federal facilities portfolio; and the availability of maintenance and repair-related data.

FEDERAL BUDGET PROCESS

Understanding the issues related to the maintenance and repair of federal facilities requires a basic understanding of how the federal budget is formulated, how maintenance and repair budget requests are compiled and reviewed, how funds are appropriated and distributed, and how expenditures are tracked.

Federal Budget Formulation

The federal budget is created through the interaction of the President, the Congress, the Office of Management and Budget (OMB), and senior managers of federal agencies. Guidelines for formulating agency budgets are developed at the highest levels of government and communicated down to the agencies. The maintenance and repair of facilities is subsumed under broader categories of agency operations and receives little, if any, specific attention at this level. However, budgets are also formulated within agencies, and requests for maintenance and repair programs, which often originate at the field office level, are then sent "up the chain" to higher levels of agency management.

Formulation of the President's Budget

Formulation of the President's budget begins around February of each year, a full year before it is released publicly and 20 months before the beginning of the fiscal year to which it applies. OMB prepares letters for executive branch offices on behalf of the President conveying the new five-year budget authority,[1] outlay estimates,[2] guidance on employment levels, requests for specific information, and other instructions related to agency budget requests. These letters are updated as necessary, and additional guidance is issued in the following months. No specific discussions on policy priorities, total budget levels, or agency allocations are held at this time.

Agencies begin developing their budgets in March or April by requesting their bureaus to develop budget options for each of their accounts. Internal agency hearings are held during the summer months. Agency budgets are traditionally submitted to OMB for review on September 1. From September through November, OMB program examiners, assisted by agency staff (when requested), review the agency estimates.

By early December, OMB completes its review, and agency heads attend a "passback" session to receive OMB's budget decisions. Agency directors are traditionally given 72 hours after the passback session to appeal. The appeals process may last through December. By law, the President is required to submit the fiscal year budget to Congress by the first Monday in February (NPR, 1993).

Congressional Review and Enactment

After submission of the President's budget, the budget committees of the U.S. House of Representatives and Senate hold hearings with agency heads, outside economists, interest groups, and others to solicit information. By April 15, Congress is supposed to produce a draft Concurrent Budget Resolution that (1) establishes the broad outlines of fiscal policy with regard to national needs, including spending and taxation; (2) sets aggregate funding levels by functional categories; (3) sets targets for total receipts and for budget authority and outlays, in total and by functional category.[3] If Congress fails to adopt a budget resolution

---

[1]Budget authority is "the authority granted to a federal agency to enter into commitments that result in immediate or future outlays. Budget authority is not necessarily the amount of money an agency or department actually will spend during a fiscal year but merely the upper limit on the amount of new spending commitments it can make" (Collender, 1995).

[2]Outlays are the "actual amount of dollars spent for a particular activity" (Collender, 1995).

[3]Functional categories are areas of general interest, including National Defense; Social Security; Income Security; Net Interest; Medicare, Health; Education, Training, Employment and Social Services; Transportation; International Affairs; General Science, Space and Technology; Agriculture; Administration of Justice; General Government; Community and Regional Development; Natural Resources and Environment; Commerce and Housing Credit; Veterans Benefits and Services; Allowances; Undistributed Offsetting Receipts; and Energy. These categories are divided into subcategories according to the major mission they fill.

by April 15, the cap included in the President's budget can be used to provide the House and Senate appropriations committees with their budgets. Once adopted, the Concurrent Budget Resolution reflects agreed upon aggregate funding levels by functional categories. Neither the budget resolution nor an accompanying report on the underlying assumptions is binding on other congressional committees. Presidential approval is not required, and the resolution is not enacted as law.

After the adoption of the Concurrent Budget Resolution, congressional authorization and appropriations committees begin drafting legislation to meet the budget resolution targets for the fiscal year. The appropriations committees must not exceed the discretionary spending cap of the Budget Enforcement Act. Appropriation bills must be signed by the President by October 1 for most federal agencies to have funds to operate. If appropriations are not passed by October 1, a continuing resolution may be passed to permit agency operations to continue at a reduced level (NPR, 1993).

Formulation of Budget Requests for Facilities Maintenance and Repair

From published reports and agency briefings, the committee learned that, over time and for a variety of reasons, federal agencies have developed individualized procedures, definitions of terms, formulas and calculations, and methodologies for developing maintenance and repair budget requests, the formulation of which generally begins at the field office or program level of each agency. For example, "Army Regulation 420-16 requires each installation to prepare an annual requirements report that specifies the installation's funding needs for operating and maintaining real property during the next fiscal year. Each major command is responsible for consolidating its installations' requirements report; then, the Assistant Chief of Engineers' Office at Army headquarters summarizes the command reports and forwards them to the Army budget office for preparation of budget requests" (GAO, 1994).

The method used to formulate maintenance and repair budget requests varies from one agency to another. Some agencies take past budgets and increase them by a certain percentage to cover inflation, new program requirements, and/or the backlog of deferred maintenance. Others target a certain percentage of the aggregate current replacement value of their agency's facilities portfolio, such as the 2 to 4 percent guideline recommended in Committing to the Cost of Ownership (NRC, 1990). Other methodologies have also been developed.

The budget request is then sent forward to higher levels of management and administration within the agency, where the maintenance and repair request is generally incorporated into the much larger "operations" or "operations and maintenance" account. This account is the principal source of funds used to pay for day-to-day expenses, such as personnel, utilities, demolition of real property, refuse handling, grounds maintenance, and other functions, as well as for facilities maintenance and repair. The Army and the Air Force's operations and maintenance account, for instance, funds many activities, such as recruiting and fielding trained units, maintaining and repairing equipment, child care and family centers, providing transportation services, civilian personnel management and pay, and maintaining the infrastructure to support the forces (GAO, 1996a). Thus maintenance and repair comprises only a small proportion of the total operations account for some, if not all, agencies.

By the time the budget request is sent from the agencies to the OMB and Congress, maintenance and repair is no longer a separate line item. The Army and the Air Force's annual operations and maintenance budget requests to Congress, for example, are presented in four broad categories: operating forces, mobilization, training and recruitment, and administrative and service-wide activities, which are further broken down into smaller categories. Except in special circumstances, little, if any, discussion takes place between federal agencies and OMB staff or congressional committees about funds to be used for facilities maintenance and repair (GAO, 1996a).

Budget Execution

Budget execution takes place during the fiscal year of October 1 to September 30, once funds have been appropriated. Funds are made available to agencies over the course of the year, not in one lump sum at the beginning of the fiscal year. This apportionment process was established by the Antideficiency Act to keep agencies from overspending early in the year, which would force Congress to pass a supplemental resolution to allow them to continue operating to the end of the fiscal year.

Agency funds must be spent on the programs, projects, and activities for which they were appropriated unless a transfer of funds has been approved or Congress is notified and consents to reprogramming, which means maintenance and repair funds may be used to meet other needs in the same appropriations category. Agencies continually review budget outlays against expenditures throughout the year and determine if reprogramming is warranted. These reviews become more frequent later in the fiscal year to ensure that spending targets are met (NPR, 1993).

When Congress appropriates the annual budget back to the agencies, the agencies themselves determine how much funding from their operations accounts will be allocated to maintenance and repair. Funds are then transferred to the field office or program level accounts. The agencies themselves also determine the projects that will be implemented.

Because maintenance and repair funds in most agencies are part of the general operations account, they are not "earmarked" for specific maintenance and repair projects. Structuring the account this way accommodates overlaps between work, operations, and alterations. For example, equipment operators often do routine equipment maintenance and alteration projects, including work that could be considered repairs (FFC, 1996). Combining operations and maintenance accounts also gives agency managers the flexibility to shift funds to operations as other, more urgent, or unanticipated needs arise or, conversely, to shift funds to maintenance and repair in case of breakdowns, to address environmental compliance issues or programmed projects, or to take care of the backlog of deferred maintenance.

The following examples illustrate why funds are shifted within the operations and maintenance account. In the first instance, in 1994, thousands of Haitian and Cuban migrants had fled their homelands for the south coast of Florida. The Navy's Atlantic Fleet was given responsibility for transporting the migrants from the south coast of Florida to Guantanamo Bay and holding them in camps. The Navy "initially found itself spending about $1 million a day . . . primarily out of its maintenance budget" to complete this mission (Peters, 1997). In contrast, for fiscal years 1993 to 1995, the Army and the Air Force obligated more funds for base maintenance and repair than were requested in their budgets even though Congress had reduced their requests. GAO speculated that this was attributable to the reprogramming of operations and maintenance funds from other accounts to base maintenance and repair (GAO, 1996a). Shifts from operating funds to maintenance and repair are most likely to occur at the end of the fiscal year when agencies have accounted for most of their operating expenses and can determine with greater certainty how much of the remaining funding can be obligated for maintenance and repair.

Accounting Structures

Accounting structures for expenditures for maintenance and repair differ from agency to agency. For example, GSA uses two accounts: (1) Operations and Maintenance and (2) Repairs and Alterations. The Operations and Maintenance account includes operations, maintenance, and minor repairs (up to a certain dollar threshold), and the Repairs and Alterations account includes all repairs, replacements, improvements, and alterations in excess of a given dollar amount with no upper limit. At the National Institutes of Health (NIH), the various institutes are assessed a certain amount each year to cover the cost of maintenance by government personnel and minor repairs by contractors. NIH also receives a direct appropriation from Congress as part of the Building and Facilities Budget to cover major repairs and improvements by contractors. The Smithsonian Institution has three categories of maintenance and repair accounts, the State Department has four, and NASA has none. NASA is funded for human space flight, science and technology, and mission support; major programs in the agency fund field installation activities, which include maintenance and repair (FFC, 1996).

Not only do accounting structures for maintenance and repair activities vary, but the definitions of elements within accounting structures also vary across agencies. As a consequence, one agency's definition of a minor alteration or of current replacement value may differ significantly from another's. Because of these variations in accounting structures and definitions, direct comparisons of maintenance and repair allocations and expenditures across federal agencies are difficult to make.

Based on briefings and other information gathered during this study, the authoring committee of this report learned that the tracking of maintenance and repair allocations and actual expenditures is not done systematically by federal agencies. This is due, in part, to the structure of the various operations and maintenance accounts and the need to shift funds during the fiscal year. Because a detailed cost accounting to determine the amount of funding actually appropriated to maintenance and repair activities is not required, few, if any, agencies complete one. Another difficulty in tracking maintenance and repair expenditures arises in the DoD agencies. "At some military facilities, maintenance and repair work is sometimes performed by uniformed personnel whose pay comes from a military personnel budget rather than an M&R budget . . . However, agencies have not developed methods for calculating the value of the contribution of such personnel or for determining the impact of such contributions on an M&R budget" (FFC, 1996).

Because of differences in methodologies, definitions, techniques to develop budgets, accounting systems, and the lack of tracking of maintenance and repair expenditures, it is difficult, if not impossible, to determine how much money and resources are allocated and spent on facilities maintenance and repair across the federal government.

Disincentives and Institutional Barriers

Effective facilities management and maintenance requires a long-term outlook and commitment; however, the annual budget process generally reinforces a short-term view. In fact, disincentives for cost effective, innovative maintenance and repair are incorporated into federal budgeting and accounting processes.

Maintenance and repair projects range from simple, inexpensive repairs to lengthy replacements of major system components that require substantial funding. In the private sector, a building owner can finance major capital projects by borrowing money and paying it back over time. "By contrast, the federal budget is a unified cash-based budget which treats outlays for capital and operating activities the same. Federal debt is undertaken for general purposes of the government rather than for specific projects or activities" (GAO, 1995a). The GAO has found that "there is a certain budget bias against capital projects, particularly when the budget is constrained, because the budget makes no distinction between an outlay for a capital asset that produces a future stream of benefits and an outlay for current operations. Because capital projects tend to require relatively large outlays in the short run, they are often foregone to meet short-term budget restraints despite their long-term benefits" (GAO, 1991). This leads to solutions that are not cost effective "because of the requirement that the entire cost of these relatively expensive assets be budgeted for in an agency's or program's annual budget or 'up-front' rather than spread over the life of the assets" (GAO, 1997a). The following examples illustrate how this budget bias can affect agency operations. At the Internal Revenue Service (IRS) building, the heating, ventilation, and air conditioning system (HVAC) could not "handle the heat generated by an expanded computer system. Consequently, IRS installed window air conditioning units to lower the temperature. Although this action avoided the capital expenditure of replacing the HVAC system . . . installation of the relatively inefficient window units increased total operating costs" (GAO, 1991). At the Pentagon, which was heated with coal-burning furnaces until 1988, the GSA, which operated the building at that time, was reluctant to replace the system because of limited resources and high replacement costs. Instead, GSA continued to repair the old system. The DoD, which occupied the building, "became so concerned about depending on the unreliable system that it began renting temporary modern boilers to heat the Pentagon at an annual cost of about $1 million" (GAO, 1991). In cases like these, it would have been more cost effective to make capital investments to operate the facilities, but the federal budget process offers no incentives to make them.

The budget process also discourages cost-effective maintenance by usually disallowing the carryover of unobligated funds from one fiscal year to the next, even if a facilities program manager can demonstrate that the most cost-effective way to implement the repair or replacement of a major operating component, such as a chiller, which may cost several hundred thousand dollars or more, may best be paid for by carrying over unobligated funds. Funds that are not spent in the current fiscal year are routinely taken back from the agencies. As an added disincentive, funding for the next fiscal year may be reduced on the premise that money not spent is money not needed. "The pressure to spend is particularly acute when the funds come from an annual appropriation, for those funds are either obligated within the fiscal year or returned to the Treasury. Subsequent year allocations may be reduced because of such returns, since failure to spend funds weakens justifications for the same level of funding the following year . . . it is currently not in any manager's interest to 'admit to' savings. It is much more rewarding to spend all your money and then claim a need for more next year than to show genuine savings" (NPR, 1993). Because there are few rewards for acting in a cost-effective, fiscally responsible manner, facilities program managers have little incentive to act in innovative ways or to take risks that may lead to more cost-effective management and maintenance programs and strategies.

Measuring the Effectiveness of Maintenance and Repair Expenditures

Determining if expenditures of maintenance and repair resources are effective is a difficult undertaking. The issue goes beyond the total dollars spent because the amount of money and resources allocated to maintenance and repair does not indicate whether those resources were used to repair mission-critical systems or to remove snow. Because government agencies do not consistently track maintenance and repair expenditures, it is difficult to develop measures to determine how effectively funds are being spent either, within or across, agencies. (For example, one measure might be total maintenance dollars spent per square foot of administrative space.) Without consistent measures, it is very difficult for facilities program managers to determine whether their maintenance and repair resources are being used optimally across their facilities inventory. Without objective benchmarks (points of reference from which measurements of any sort may be made) by which to identify "best practices" among the agencies, information that could be shared and used across agencies to improve government performance in this area is not available.

Legislative Requirements

Pressures on already limited maintenance and repair budgets may be increased through legislative requirements to improve health, safety, or welfare that have facilities-related impacts. The purpose of these requirements may be laudable, but they are usually enacted without the funding to implement them. These so-called "unfunded mandates" result in de facto reductions in agencies' operations and maintenance budgets by requiring them to make additional repairs and alterations using current appropriations, thus dividing up the maintenance and repair "pie" into smaller and smaller pieces.

The Americans with Disabilities Act of 1990, for example, prohibits discrimination against people with disabilities in employment, transportation, public accommodations, communications, and activities of state and local government. The law requires the removal of barriers from existing facilities, if this is readily achievable, and requires making the altered facilities as accessible as is feasible. However, funds for removing barriers and/or improving accessibility were not appropriated. For most federal agencies, the funding has to come from current operations and maintenance budgets. Similarly, meeting other unfunded mandates, such as removing hazardous materials, must also be paid for from already constrained agency operations and maintenance accounts.

Data on the exact costs of complying with these legislative requirements are not available. However, anecdotal information reported by the GAO indicates that these costs are substantial and have had an impact on operations and maintenance budgets. The GSA, for example, spent "about $40 million to remove asbestos from a federal building in San Francisco . . . [and] in a federal building in Denver . . . the cost of installing a sprinkler system increased by about $1.5 million after asbestos was discovered in the ceiling" (GAO, 1991). At Glacier National Park, "federal requirements for lead paint abatement, asbestos removal, surface and waste water treatment, and accessibility for disabled visitors required park managers to divert staff time and operating funds from visitor and resource management activities" (GAO, 1995b).

FEDERAL FACILITIES PORTFOLIO

In today's dynamic federal environment, agencies' changing needs for workspace, facilities, and technology have significant implications for maintenance and repair. With reduction of the federal workforce, the elimination of programs, and changes in agencies' missions, federal employees are being moved from headquarters to field offices, from one field office to another, or from field offices to headquarters, to improve program efficiencies and reduce costs, among other objectives. These changes have resulted in underutilized space in some locations and in overutilized or "overpopulated" space in other locations. As agencies' missions change, some facilities are no longer used at all. The most dramatic example, but not the only example, is the Base Realignment and Closure process, which has resulted in the closure of one of every five military installations across the country. The elimination of 107,000 civilian full-time positions through downsizing could result in millions of square feet of federal office space becoming unneeded or underutilized (GAO, 1997b).

Underutilized Facilities

As federal agencies eliminate staff positions, so-called "reductions in force" do not occur across the board but are targeted to specific functions or programs. Thus, an agency program that was once staffed at a level requiring five floors in a building, may now be staffed at a level requiring the occupancy of two floors. This 60 percent reduction in space needs does not, however, translate into a proportionate reduction in maintenance and repair needs or costs. As long as the building is occupied, 100 percent of a broken furnace or air conditioning system has to be repaired. Similarly, maintenance and repair costs cannot be reduced in proportion to reductions in the number of employees. The integration of critical operating components (heating, plumbing, ventilation, electrical, fire, communications, and safety systems) usually requires that entire systems be maintained in good working order to protect workers' health, safety, welfare, and productivity.

Even closing complexes or installations does not necessarily mean that they no longer require maintenance and repair. For example, some military bases have been closed, but DoD has retained ownership while local communities have sought ways to reuse the facilities. In the interim, DoD is responsible for protecting and maintaining these bases. It has been estimated that the "overall cost to maintain bases closed in the 1988 and 1991 rounds was approximately $290 million through fiscal year 1996." In general "maintenance levels have not been reduced from their initial levels, even where progress toward reuse has been slow" because DoD has attempted to keep the facilities in a reusable condition in response to the demands of surrounding communities and political pressure to support the communities affected by base closures (GAO, 1997c).

If federal facilities have unique or specialized functions, other factors come into play in cutting the costs of operations and maintenance in underutilized facilities. The Plum Brook Station of NASA's Lewis Research Center, for instance,

operates on a cost-reimbursable basis, with most of its operating cost covered by revenue from users of four test facilities at the station. Even if all four of the test facilities were closed, the operating cost would still be about $2 million, primarily because the Nuclear Regulatory Commission requires that the reactor be maintained in its current state. The only way to close the location and dispose of the property would be to dismantle the reactor. However, the cost for doing this would be prohibitively expensive—about $100 million in 1997 dollars, according to a 1990 estimate. In addition, there are no disposal sites to accommodate the radioactive waste that would be generated by the dismantling process (GAO, 1996b).

Overutilized or Overpopulated Facilities

As agencies attempt to reduce their costs of ownership, office buildings are being retrofitted to accommodate more people, more computers, and more support equipment in the same amount of space. More intensive use of building components, such as elevators, plumbing, and electrical systems, increases the rate of wear and tear on building components and increases the need for maintenance and repair. Facilities managers at the IRS, the U.S. Customs Service, and the Federal Aviation Administration have reported that they have to "constantly 'jerry-build' electrical systems to accommodate expanding computer needs because the older electrical systems in the buildings cannot handle the demands . . . more computers place additional strains on the air conditioning systems, not only to cool mainframe computers but also to counteract the heat generated by desktop units" (GAO, 1991). Thus, more intensive use of facilities can place additional burdens on building systems and components and increase the need for maintenance and repair.

Excess Facilities and Changing Missions

The federal facilities portfolio has grown over time in response to new programs and requirements, defense and foreign policy initiatives, changing demographics, and other factors. More than 500,000 facilities are currently owned by the federal government. In 1995, it was estimated that federal civilian agencies alone occupied more than 750 million square feet of office space in thousands of government-owned and government-leased buildings nationwide (GAO, 1997b). The Federal Property Management Regulation "requires agencies to conduct an annual review of real property to ensure prompt identification and release of unneeded or underutilized property" and to "maintain the minimum inventory necessary to conduct its mission" (GAO, 1997c). In practice, little emphasis has been placed on demolishing obsolete facilities or divesting the government of unneeded, but still viable, properties.

Based on the information available to the committee, it appears that the number of excess facilities is increasing throughout the federal government as agencies realign their missions in response to changing circumstances. The most dramatic example is the Base Realignment and Closure process. Between 1988 and 1995, four "rounds" of domestic military base closings were implemented as part of the U.S. military's restructuring of its mission in the post-Cold War era. Approximately one in every five military installations in the United States is slated to close by 2001, creating thousands of "excess" federal military buildings and other constructed facilities. Some civilian agencies also report having more facilities than they need to meet their current and anticipated mission requirements. The Atomic Energy Commission's (now the U.S. Department of Energy's [DOE's]) mission after World War II was to build a nuclear arsenal. Today, DOE is faced with addressing the resulting environmental, health, and safety risks at thousands of contaminated sites. DOE's Strategic Plan includes a strategy to "complete about 100 surplus nuclear facility deactivation's during FY 1998 and FY 1999. This is about 10 percent of the total remaining facilities that require deactivation." As of October 1995, the "Department of State had identified over 100 overseas properties valued at $467 million for potential sale" (GAO, 1996c).

Other agencies are also likely to have excess facilities as their missions change. A case in point is the Department of Veterans Affairs (VA). The VA Health Care System was "established in 1930, primarily to provide for the rehabilitation and continuing care of veterans injured during wartime service." The VA provided direct care to its clients and owned and operated its own health care facilities, becoming the country's largest direct delivery system (GAO, 1996d). In recent decades, the size of the VA's client population has been declining. As the remaining population ages, its needs for health care services are changing. Between fiscal years 1980 and 1995, the days of hospital care fell from 26 million to 14.7 million, but the "number of outpatient visits increased from 15.8 million to 26.5 million; and the average number of veterans receiving nursing home care in VA-owned facilities increased from 7,933 to 13,569" (GAO, 1996d). In response to these and other trends in the health care field, the VA is changing its mission from providing in-patient services to providing health care services on an out-patient basis, sometimes through partnerships with private sector health care providers. As a result, "about 50,000 VA hospital beds were closed or converted to other uses between 1969 and 1994, and further declines are likely in the next 7–10 years" (GAO, 1996d). Underutilized facilities are being created in this transition, but as of March 1996, the VA had "not closed any hospitals because of declining utilization (GAO, 1996d)." [4] Like other federal agencies undergoing

---

[4]Two hospitals were closed when they were damaged by earthquakes.

similar mission realignments, the VA is still responsible for maintaining and re-pairing its underutilized facilities.

Changing demographics and social trends do not always result in excess fa-cilities, and not all federal agencies own excess facilities. For example, to re-spond to increased caseloads and legislative changes regarding the sentencing of criminals, additional courthouses, prisons, and related facilities are being built nationwide. Not enough information was available for the committee to deter-mine precisely how many federal agencies own excess facilities. Nevertheless, there is evidence that some agencies own buildings and infrastructure that are no longer needed or are otherwise underutilized.

The demolition of excess facilities would require an up-front investment of funds but could, in the long run, be cost effective through annual savings on operations, utilities, and maintenance and repair funds. The demolition of federal facilities, however, can be expensive, time-consuming, and difficult. The military services estimate that demolition costs for facilities other than World War II era wooden barracks range from $8 to $12 per square foot (GAO, 1997d). The mag-nitude of the costs for demolishing excess federal facilities can be illustrated by the U.S. Army's demolition program. The Army "has the largest and most cen-trally directed demolition program" among DoD agencies. "Starting in fiscal year 1998, the service plans to spend $100 million in O&M [operations and mainte-nance] funds per year through fiscal year 2003 to eliminate excess space at an estimated cost of $10 per square foot." GAO calculates that it will take the Army about 13 years to eliminate its excess space at a cost of about $1.3 billion (GAO, 1997d). The total cost of demolishing all excess federal facilities would be sub-stantially higher, and many older buildings would require special procedures for the removal of asbestos, underground storage tanks, and other items.

The demolition of excess federal buildings might also be de facto reductions in maintenance and repair budgets in the short term. Demolition, like mainte-nance and repair, is funded from the operations budget in most agencies. Unless special appropriations for building demolition have been approved by Congress, substantial demolition projects could cut into the funds that might otherwise be used for the maintenance and repair of occupied facilities. To program managers faced with the choice of demolishing an unused facility or repairing an occupied one, demolition is a much less viable option.

Agencies may also try to reduce the number of excess facilities by declaring them surplus facilities and allowing other federal agencies or nonfederal entities to compete for their use. Once it has been determined that no other federal agency has a use for a facility, the agency can offer it to state and local agencies and then to the general public. However, because of security requirements, the locations of buildings within larger installations or complexes, or building design and con-struction factors, such as central utility infrastructure, it is not always possible or desirable to transfer title to a nonfederal owner. Property transfers that are fea-sible may take years to complete.

Transferring the title of a facility brings with it the responsibility to meet environmental regulations, which may be costly. In reviewing possible consolidations and closures at NASA and estimates of cleanup costs, "officials at several centers believed the cost could be as much as two to five times higher than if NASA were to retain the property. The higher cost would occur if NASA cleaned up facilities to meet more stringent standards that might be required for disposal" (GAO, 1996b). Transferring ownership, therefore, may not always be cost effective, particularly if significant funds are required to meet environmental or other regulations prior to the transfer. The disposition of former nuclear sites and their associated facilities present unique situations that are not easily resolvable through any of the aforementioned strategies.

New Construction

Concurrent with the creation of excess facilities, the federal government continues to build new facilities in response to changing needs. Federal courthouses and prisons are being built throughout the country to meet the needs of the criminal justice system. Embassies are being built in newly established countries in Eastern Europe and Africa to support foreign policy and serve U.S. citizens abroad. Substantial new military facilities are being built as a result of the Base Realignment and Closure process; as military divisions and functions are moved from bases slated for closure to active ones, new facilities are being constructed to accommodate the transferred personnel and functions.

The acquisition of new facilities has an impact on operations and maintenance budgets, although it is not readily apparent. Because the budget process is structured to focus on the first costs of design and construction rather than the life-cycle costs of the facility, the costs of design and construction for substantial new facilities are usually paid from specific line items in the budget. However, as new facilities come on line, their operation and maintenance costs must be funded from current operations and maintenance accounts. The potential impact of new facilities on an agency's existing operations and maintenance programs does not have to be analyzed in the budget review process. Given already constrained budgets, new facilities can reduce the funds available for maintaining existing buildings. As new facilities age, the requirements for repairs increase, which increases their overall impact on the operations and maintenance budget.

AVAILABILITY OF MAINTENANCE AND
REPAIR-RELATED DATA

Throughout this study, the committee was hampered by the lack of data regarding maintenance and repair needs and expenditures. Facilities program managers are similarly hampered by a lack of research and information on maintenance and repair-related activities. Information regarding the maintenance and repair-related aspects of facilities management is not readily available because "[r]esearch work in facility management, especially cost prediction of facilities, has only begun recently" (Christian and Pandeya, 1997). Managing the Facilities Portfolio, a report of a study conducted in 1991, found that "though much has been written about the causes and size of the deferred maintenance problem, comparatively little has been published on models for analyzing, assessing, and funding facility assets on a consistent basis" (AME, 1991). For day-to-day maintenance decisions, no standard methods have been established for diagnosing problems or identifying and evaluating appropriate treatments (Shen and Grivas, 1996). And, "a major cost analysis problem is that of identifying the lowest cost repair option to repair a component, especially where alternative repair options are available and the costs differ substantially" (Urban Institute, 1994). The lack of research-based information significantly hinders facilities program managers in developing cost-effective facilities management and maintenance programs and in making effective arguments to justify budget requests.

This is not to say that no maintenance and repair-related data exist. A number of private companies and nonprofit organizations regularly gather and publish data on facility maintenance and repair costs and operations. Reports published by the R.S. Means Company, Incorporated, the Whitestone Companies, the Association of Higher Education Facilities Officers, the International Facilities Management Association, and the Building Owners and Managers Association International, among others, include facilities cost data indices, such as per-square-foot costs for building operations and maintenance and standardized charts of accounts for private-sector office buildings and other commercial facilities. Most of this information, however, relies heavily on data gathered from the private sector or academic institutions, and some state and local governments, but not federal facilities. Furthermore, most cost estimating guidelines are based on construction and major renovations; few, if any, are specifically designed to cover a wide range of maintenance and repair items (AME, 1991).

Three important facilities management areas about which relatively little research has been done are: (1) the relationship between timely maintenance and the avoidance of future costs; (2) the deterioration/failure rates of building components; and (3) the nonquantitative implications of building maintenance (or lack thereof).

Timely Maintenance and Future Cost Avoidance

When agency managers are asked to justify their maintenance and repair budget requests, "public officials . . . find the most convincing and compelling information to be the future costs that can be avoided by undertaking early, preventive or corrective maintenance activities" (Urban Institute, 1994). But "there appear to be no standard methods for estimating how much additional future cost will be incurred by a decision in the present to defer maintenance on a particular facility" (Urban Institute, 1994). Although some federal agencies account for and track backlogs of deferred maintenance and repair projects, the committee was not aware of any analyses that have been done to identify the effects of continuous backlogging or the failure to perform adequate maintenance and repair.

Very little research has been published on cost avoidance in relation to the maintenance and repair of building components. "Good information does not seem to be available as to what will happen to components if they are not repaired soon, e.g., how fast a small leak will develop into a much larger and much more costly to repair, leak" (Urban Institute, 1994). This type of information would help facilities program managers plan and implement cost-effective maintenance and repair programs. It would allow them to anticipate the optimal time to repair or replace an operating component to take advantage of its maximum service life, but replacing it before it deteriorates to the point of disrupting business or otherwise affecting an agency's mission. Research-derived, cost avoidance information that is meaningful to public officials and senior decision makers would also help facilities program managers make more compelling arguments for adequate maintenance and repair funding.

Deterioration/Failure Rates of Building Components

Relatively little research has been done on the deterioration rates of building components. "The state-of-the-art in predicting deterioration/failure rates for infrastructure is not well-developed . . . . A small amount of work has been done to develop deterioration rates for various types of sewer pipe and very little, thus far, has been done for buildings" (Urban Institute, 1994). The U.S. Army's Construction Engineering Research Laboratories have made a significant effort to predict deterioration rates as a component of their Engineered Management Systems (EMS) programs. These systems, which are discussed in greater detail in Chapter 3, include estimates of deterioration rates for specific types of infrastructure (e.g., road pavement, gas lines, and roofs) in order to develop optimal replacement strategies that indicate when repairs should be done and at what cost. The newest EMS program, BUILDER, incorporates "deterioration curves developed from experience over time" (Uzarski and Finney, 1997). Only field experience with the BUILDER program will show how useful and accurate it is.

The general lack of research on the deterioration of building components is due, in part, to the complexity of the issue. The rate of deterioration of building components is affected by a number of factors, including but not limited to, the quality of construction and materials, the extent of use, the level of maintenance, weather conditions, soil conditions, structural loading, and environmental pollutants. This large number of variables makes it difficult to construct accurate models of component deterioration. Appropriately structured studies could yield valuable information about the costs, benefits, and other effects of timely maintenance on the deterioration of building components, and, conversely, on the financial and other implications of the lack of maintenance on building component service life. Predictions of deterioration/failure rates could be useful for estimating future budget needs and determining optimal repair/replacement cycles for particular types of infrastructure (Urban Institute, 1994). The results would also help for analyzing the life-cycle of maintenance and repairs (Uddin and Najafi, 1997).

Nonquantitative Effects of Maintenance

Winston Churchill, addressing the House of Commons, once said, "We shape our buildings, and afterwards our buildings shape us." Although "procedures for assessing the nonquantitative impacts of poorly maintained buildings, thus far, appear to be based primarily on qualitative, subjective judgments rather than on empirical data and analysis" (Urban Institute, 1994), mounting evidence suggests that not only can the design of buildings affect workers' productivity, but the level of maintenance and repair can also affect building users. Faulty wiring, leaking roofs, and deteriorating pipes can lead to equipment failures and other business disruptions. Defects can cause discomfort, safety, or health risks for occupants. Deferred maintenance also "contributes to poor quality working space, impedes agencies' operations, and in some instances, jeopardizes employees' health and safety" (GAO, 1991). Evidence suggests that the quality of the work environment, including the quality of the facilities, can affect employee morale and is a factor in recruiting and retaining workers. It has been estimated that over the 40-year life of a typical office building, the salary costs of the people working in that building will equal 95 percent of the total costs of constructing, operating, staffing, and maintaining it (FFC, 1997).

The environment and performance of buildings can contribute to illnesses and injuries. Health effects associated with the indoor environmental quality of buildings can be categorized into building-related illnesses and "sick-building syndrome." Building-related illnesses, medical conditions with known etiologies, such as Legionnaire's disease, respiratory allergies, or asthma, can often be traced to specific sources and eliminated. "Appropriate building design, operation and maintenance procedures can prevent such illness in the first place" (NSTC, 1995).

Sick-building syndrome is characterized by nonspecific symptoms and less well defined health problems. Current evidence suggests that improved building design, operation, and maintenance can reduce a variety of still unidentified indoor contaminants and thereby prevent or reduce symptoms among building occupants (NSTC, 1995).

Lost workdays attributable to building-related health problems decrease productivity and could affect an agency's mission-related activities. If worker absenteeism due to building-related factors can be reduced, productivity should increase. Worker compensation for lost time, illnesses, and lawsuits can also have a significant effect on an agency's productivity and budget.

REFERENCES

AME (Applied Management Engineering). 1991. Managing the Facilities Portfolio: A Practical Approach to Institutional Facility Renewal and Deferred Maintenance. Washington, DC: National Association of College and University Business Officers.

Christian, J., and A. Pandeya. 1997. Cost predictions of facilities, Journal of Management in Engineering 13(1): 52–61.

Collender, S. 1995. The Guide to the Federal Budget, Fiscal Year 1996. Washington, D.C.: The Urban Institute Press.

FFC (Federal Facilities Council). 1996. Budgeting for Facilities Maintenance and Repair, Technical Report No. 131. Washington, D.C.: National Academy Press.

FFC. 1997. Federal Facilities Beyond the 1990s: Ensuring Quality in an Era of Limited Resources, Technical Report No. 133. Washington, D.C.: National Academy Press.

GAO (General Accounting Office). 1991. Federal Buildings: Actions Needed to Prevent Further Deterioration and Obsolescence. Report to the Chairman, Subcommittee on Public Works and Transportation, U.S. House of Representatives. GGD-91-57. Washington, D.C.: Government Printing Office.

GAO. 1994. Financial Management: Army Real Property Accounting and Reporting Weaknesses Impede Management Decision-Making. Letter Report. AIMD-94-9. Washington, D.C.: Government Printing Office.

GAO. 1995a. Budget Issues: The Role of Depreciation in Budgeting for Certain Federal Investments. AIMD-95-34. Washington, D.C.: Government Printing Office.

GAO. 1995b. National Parks: Difficult Choices Need to be Made About the Future of the Parks. Chapter Report. RCED-95-238. Washington, D.C.: Government Printing Office.

GAO. 1996a. Operation and Maintenance Funding: Trends in Army and Air Force Use of Funds for Combat Forces and Infrastructure. Report to the Chairman, Subcommittee on National Security, Committee on Appropriations, U.S. House of Representatives. NSIAD-96-141. Washington, D.C.: Government Printing Office.

GAO. 1996b. NASA Infrastructure: Challenges to Achieving Reductions and Efficiencies. Report to the Chair, Subcommittee on National Security, International Affairs, and Criminal Justice, Committee on Government Reform and Oversight, U.S. House of Representatives. NSIAD-96-187. Washington, D.C.: Government Printing Office.

GAO. 1996c. State Department: Millions of Dollars Could Be Generated by Selling Unneeded Real Estate Overseas. Testimony. NSIAD-96-195. Washington, D.C.: Government Printing Office.

GAO. 1996d. VA Health Care: Opportunities to Increase Efficiency and Reduce Resource Needs. Testimony. T-HEHS-96-99. Washington, D.C.: Government Printing Office.

GAO. 1997a. Budgeting Issues: Budgeting for Federal Capital. Chapter Report. Report to the Chair, Committee on Government Reform and Oversight, U.S. House of Representatives. AIMD-97-5. Washington, D.C.: Government Printing Office.

GAO. 1997b. General Services Administration: Downsizing and Federal Office Space. Testimony. T-GGD-97-94. Washington, D.C.: Government Printing Office.

GAO. 1997c. Military Bases: Cost to Maintain Inactive Ammunition Plants and Closed Bases Could Be Reduced. Letter Report. NSIAD-97-56. Washington, D.C.: Government Printing Office.

GAO. 1997d. Defense Infrastructure: Demolition of Unneeded Buildings Can Help Avoid Operating Costs. Report to the Chair, Subcommittee on Military Installations and Facilities, Committee on National Security, U.S. House of Representatives. NSIAD-97-125. Washington, D.C.: Government Printing Office.

NPR (National Performance Review). 1993. From Red Tape to Results: Creating a Government That Works Better and Costs Less: Mission-Driven, Results-Oriented Budgeting. Washington, D.C.: Government Printing Office.

NRC (National Research Council). 1990. Committing to the Cost of Ownership: Maintenance and Repair of Public Buildings. Building Research Board, National Research Council. Washington, D.C.: National Academy Press.

NSTC (National Science and Technology Council). 1995. National Planning for Construction and Building R&D. Washington, D.C.: Government Printing Office, NISTIR 5759.

Peters, K. 1997. Funding the Fleet. Government Executive 29(1): 42–45.

Shen, Y. and D. Grivas. 1996. Decision-support system for infrastructure preservation, Journal of Computing in Civil Engineering. 10(1): 40–49.

Uddin, W. and F. Najafi. 1997. Deterioration Mechanisms and Non-Destructive Evaluation for Infrastructure Life-Cycle Analysis. Pp. 524–533 in Infrastructure Condition Assessment: Art, Science, and Practice. Mitsuru Saito, ed. New York: American Society of Civil Engineers.

Urban Institute. 1994. Issues in Deferred Maintenance. Washington, D.C.: U.S. Army Corps of Engineers.

Uzarski, D. and D. Finney. 1997. BUILDER-Managing Buildings. The Military Engineer. 89(586): 36–37.

3

Condition Assessments

To effectively manage its facilities portfolio, an organization must establish meaningful baseline data about the size and physical condition of its facilities. This information is used to estimate short- and long-range maintenance and repair needs. Many organizations with facilities portfolios have established systematic condition assessment programs to provide this baseline management information. A relatively small number of organizations uses these data as part of an integrated capital assets management program.

Condition assessments and a capital assets management program are key components of an effective maintenance and repair program. Tasks 2 and 3 of the committee's charge were to investigate the role of technology in performing automated condition assessments and to identify the staff capabilities necessary to perform condition assessments and develop maintenance and repair budgets. This chapter reviews the components of condition assessments and capital assets management programs; describes the use of condition assessments by federal agencies; reviews technologies for automating condition assessments; identifies the staffing implications for performing automated condition assessments; and identifies issues related to condition assessment programs as they are currently implemented by federal agencies.

COMPONENTS OF A CONDITION ASSESSMENT AND CAPITAL ASSETS MANAGEMENT PROGRAM

A condition assessment has been defined as the "process of systematically evaluating an organization's capital assets in order to project repair, renewal, or replacement needs that will preserve their ability to support the mission or activities they were assigned to serve" (Rugless, 1993). Condition assessment programs generally begin with inspections of individual facilities by trained personnel who can determine the physical condition and functional performance of facilities, as well as maintenance and repair deficiencies. Building systems, components, and materials are inspected for outright signs of deterioration or failure, as well as for more subtle symptoms indicating abnormal conditions (NRC, 1990).

Information gathered from inspections can be used to (1) estimate the need for maintenance and repair, (2) develop cost estimates and funding priorities for various projects, and (3) generate and prioritize work orders. Facility condition assessments can also be used to: evaluate deferred maintenance and funding requirements; plan a deferred maintenance reduction program; compare conditions between facilities; establish baselines for setting goals and tracking progress; provide accurate and supportable information for planning and justifying budgets; facilitate the establishment of funding priorities; and develop budget and funding analyses and strategies (AME, 1991).

Studies by the National Research Council, the Federal Facilities Council, and the GAO have reported that condition assessment programs are used by some federal agencies to identify and document maintenance and repair backlogs or deferred maintenance. Increasingly, computer software programs and other technologies are being used to support the gathering and analysis of condition assessment data as part of a capital assets management program (CAMP). "Together, a condition assessment survey and capital assets management program (CAS/CAMP) can provide a process for establishing condition standards, and inspecting, recording, reporting, prioritizing, and managing the maintenance and repair of buildings and infrastructure." They can also be used as a "tool to construct cost reports, develop priorities, and create life cycle analyses as well as provide the ability to construct a wide variety of ad hoc reports" (Rugless, 1993).

A CAS/CAMP, as defined by the authoring committee, incorporates four key components:

- a standardized, documented inspection process that provides accurate, consistent, and repeatable results
- a detailed, ongoing inspection of real property assets, including facilities, infrastructure, and large, in-place equipment that is validated at predetermined intervals
- standardized cost data based on an industry-accepted cost estimating system to determine repair and replacement costs
- a user-friendly information management system or process that prioritizes current and anticipated maintenance and repair requirements to maximize the utilization of resources (labor and dollar) and return on investment (ROI)[1] and minimize the cost of irreversible loss of service life[2] and total penalty cost.[3]

---

[1]The ROI calculation "stems basically from comparing the extended life of the component or system that is being repaired or replaced to the remaining life of the existing item should no repair be

Condition Assessment Surveys

The CAS provides the evaluation criteria for inspecting each facility and is intended to ensure consistency among inspectors, assets, inspection intervals, and geographical locations. Data collection can be standardized through automated checklists and/or written guidelines to ensure that data are consistent from one facility to another; data can also be "rolled up" to represent larger numbers of buildings. Standardized inspection processes range from macrolevel assessments of facility systems for organizations with work-order systems in place to detailed microlevel inspections of individual buildings that identify deficiencies and can be used for managing daily work orders. A standardized inspection process with uniform deficiency standards and inspection methods can enable a facilities program manager to compare and prioritize inspection data from different facilities, equipment, infrastructure systems, locations, and inspectors.

Standardization of inspection data is important because "human inspectors make subjective judgments, and their measurements and ratings are highly variable", whereas automated systems are less subjective and more consistent (Sanford and McNeil, 1997). In an effort to provide for greater consistency and objectivity, some organizations have developed standards or guidelines for inspections that may or may not be automated. The U.S. Army's Construction Engineering Research Laboratories, for instance, has developed standardized condition indexes for inspections. "Utilizing sampling techniques, inspectors view the building at the component level looking for which, if any, of 20 'generic' distresses and their severities are present. Point values are assigned to each distress type/severity/density combination found. These are summed, corrected, and subtracted from 100 to obtain a Building Component Condition Index" (Uzarski and Burley, 1997).

Standardized checklists cannot, in and of themselves, provide for consistency in condition assessments across facility inventories. Buildings and their component systems are complex, and the identification of the causes of building deficiencies requires that inspectors be trained to recognize the root causes of deficiencies, not just the visual manifestations of problems. For example, peeling paint on a wall or ceiling may be caused by poor quality paint or may indicate a much more serious problem, such as a leaking roof. Inspectors should be trained to recognize these differences, to develop alternative courses of corrective action,

---

made." To calculate the ROI, the repair cost is amortized over the lengthened life of the component or system and compared to the amortized cost of replacing the item earlier if the repairs are not made. If a repair would greatly lengthen the remaining life of the component, a favorable ROI is likely to result (Rugless, 1993).

[2]Cost of irreversible loss of service life can be defined as the cost associated with the loss of remaining service life if repairs are deferred.

[3]Total penalty cost can be defined as the total cost of deferred repairs.

to cost out the alternatives, and to determine the cost effectiveness of making a repair in the short term, the long term, or not at all.[4]

Capital Assets Management Programs

The CAMP, in contrast to the CAS, is a systematic approach to scheduling and budgeting current and anticipated deficiencies that maximizes the ROI and preserves the value of the physical asset. The prioritization process incorporates both quantitative data (the condition of the inspection item or overall system, irreversible loss of service life cost, and total penalty cost) and qualitative data (importance of building function and mission). As parameters change, the data can be revised or updated because the CAMP is a "living document."

The CAMP, combined with the knowledge of trained inspectors and facilities program managers, can also be the basis for making repair-versus-replace decisions, identifying maintenance and repairs that can be deferred without loss of investment, and projecting long-range capital renewal requirements. The CAMP can provide a standardized, cost-effective, convenient approach to:

- establishing a systematic and economical method for periodically updating asset conditions
- minimizing surprise failures of equipment or systems
- determining critical maintenance requirements for physical assets
- optimizing operations and maintenance dollars to maximize the ROI for large assets
- developing sound, defensible budgets based on an organization's goals and objectives rather than on the physical condition of facilities alone
- providing a consistent methodology for comparing requests from facility managers
- allocating maintenance and repair dollars among competing requirements or organizations

The application of a CAS/CAMP by federal facilities program managers can be broken down into four broad categories:

- CAS as a tool for identifying and validating deferred maintenance backlog
- CAS as a means of evaluating the condition of physical assets and their maintenance against projected life cycles
- CAS/CAMP as a process for identifying, prioritizing, and managing the asset portfolio
- CAMP as a decision-making tool for trade-off analyses for allocating resources

---

[4]In some circumstances, it may not be cost effective to make early inspections and repairs but only to make corrections when needed. For instance, using remote control video cameras to inspect the interior condition of miles of underground pipes may be more expensive and less cost effective than simply replacing a pipe once it breaks, depending on where the pipe is located, the type of business disruptions, and the personal or real property damage the breakage causes.

Together, the CAS/CAMP would help facilities program managers identify, prioritize, and manage the overall condition of their physical inventories and provide a sound, defensible tool for articulating the "business case" for investing resources in infrastructure.

CAS/CAMP covers a wide range of methodologies, levels of detail, automated support systems, and budgeting objectives. Although many systems and software programs are already available, the industry is still in the development stage. Few standards have been established, and the technology is changing rapidly. An internal organizational evaluation to determine the objectives and reporting requirements of the facilities management program is essential for federal program managers when choosing an appropriate system.

USE OF CONDITION ASSESSMENTS BY FEDERAL AGENCIES

Condition assessment survey programs were first used by federal agencies, including the U.S. Air Force Strategic Air Command, the U.S. Navy, the U.S. Army, and the U.S. Department of Energy, in the 1960s and mid-1970s. These programs were mostly focused on identifying backlogs of maintenance and repair or deferred maintenance and projecting future funding requirements for facilities and infrastructures.[5]

In the 1980s, more sophisticated inspection and engineering management programs were developed, such as the U.S. Army Construction Engineering Research Laboratories' suite of engineered management systems (EMS), including PAVER, RAILER, BRIDGER, ROOFER, GPIPER, and others. These computerized systems use condition assessment and prediction techniques to develop annual and long-range work plans based on timely maintenance and repair using specified policies and strategies. The EMS programs are designed to provide objective, repeatable methods for inspections and evaluations, work histories of facilities, information for real property updates, the consequences of decision alternatives, and quick access to engineering technology. All of these systems, however, were designed for individual infrastructure elements or building components, not for complete buildings. Thus, PAVER focuses on pavement management, RAILER on railroad tracks, BRIDGER on bridges, ROOFER on roofs, and GPIPER on underground gas pipes. The committee did not have adequate information to determine whether or under what circumstances the EMS programs are cost effective.

In the early 1990s, the DOE, and later the DoD, undertook expansive programs to develop and implement standardized CAS programs across their entire infrastructures. Both departments focused on developing comprehensive processes that included detailed inspection standards, inspector training programs,

---

[5]Representative programs include the Strategic Air Command's Pavement Condition Index, the U.S. Army's IFS-1, and the U.S. Navy's Annual Inspection Summary.

automated data collection devices, and the ability to aggregate information at multiple levels based on location and organization. The DOE's CAS was "initially conceived in 1990 as an industry-based system of standards to develop deficiency-based capital maintenance and repair costs for use in managing DOE real property assets." The DOE developed basic standards and automated systems, field implementation and maintenance, and enhancement and field support (USACIR, 1996). DOE's goals for the program were to "have available for all of the field offices and contractors, a system that was simple and easy to use; achieve[d] results that were as accurate and timely as possible; and report[ed] results consistently from site to site and across programs" (Earl, 1997). CAS manuals were developed outlining deficiency standards and inspection methods systematically. Nationally recognized, geographically adjustable cost algorithms were used, as well as an automated system of cost collection, to integrate and summarize building level detail to a site-wide level. The use of state-of-the-art tools by inspectors, such as bar coding and hand-held computers, was encouraged (Earl, 1997). However, the initial concept of "rolling up" the data to produce a headquarters-level report was dropped early in the program, which left DOE operations offices and their secretarial offices on their own in deciding the type of CAS system they would employ. Since 1990, the DOE system has been revised, the software upgraded, and functional improvements made (USACIR, 1996).

In the Defense Appropriations Act for fiscal year 1992, funds were provided for the implementation of a pilot test program to conduct comprehensive maintenance surveys, referred to collectively as the condition assessment survey, at selected DoD installations. The pilot test of the surveys was completed in 1995. The pilot information system could calculate costs for repairs and replacements and rate facilities on a condition index scale of 0 (poor) to 100 (excellent). This computerized system could also determine the time frame for repairs based on a facility's ROI. However, DoD projected the cost of implementing the system across all services would be about $715 million, and service officials reported that the system's information management system was labor intensive and expensive to maintain. For these reasons, the system was never deployed (GAO, 1997).

The Air Force Commanders' Facility Assessment (CFA) program was initiated in 1992. The CFA is "designed to link facility condition to mission requirements to ensure that resources for maintenance, repair, and minor construction are allocated to the most critical mission needs of field commanders" (GAO, 1997). The assessments are intended to help commanders stratify their maintenance and repair and military construction requirements for real property. Under this program, field commanders can identify recurring (day-to-day requirements) and nonrecurring requirements. Nonrecurring requirements are classified as Levels I, II, or III. Level I (unsatisfactory) facilities have deficiencies that cause frequent mission interruptions, accelerate the deterioration of the facility, result in high life-cycle maintenance costs, curtail or eliminate some operations, or degrade livability and workplace conditions. Level II (degraded) facilities have deficiencies that impair mission support, reduce the effectiveness of the workforce, or accelerate the deterioration of the facility. Level III (adequate) facilities are in good enough condition that they do not impair accomplishment of the mission, although they may have some minor deficiencies. The program provides condition data for Air Force facilities for fiscal years 1993 and 1995. Improvements in the program and enhancements to the software are under way (GAO, 1997).

The U.S. Army began using its Installation Status Report (ISR) system in 1995 to "assist installations in articulating their infrastructure needs to the DA [Department of the Army] and to allow the DA to develop funding requests for Congress" (O'Hara et al., 1997). The ISR established department-wide standards for each type of building on Army installations. Based on annual inspections, estimates of funding to sustain and/or renovate facilities based on national averages for military construction were developed. The ISR also calculates the funding needed for new construction to ensure that each installation has the facilities necessary to fulfilling its mission (O'Hara et al., 1997). The data for fiscal years 1995 and 1996 on the status of installations are grouped into five broad areas: mission facilities, strategic mobility facilities, housing facilities, community facilities, and installation support. The ISR also includes ratings indicating whether the Army has enough facilities and whether existing facilities meet Army standards (GAO, 1997). Refinements to this program are planned.

A new EMS being deployed by the Army's Construction Engineering Research Laboratories, called BUILDER, is designed for owners of large numbers of buildings for managing building assets, both individually and in groups, including the development of long-range maintenance and repair plans. BUILDER "combines inventory, inspection, condition assessment, condition prediction, and M&R [maintenance and repair] planning features in a Windows™ software environment" (Uzarski and Burley, 1997). The BUILDER system incorporates a variety of technologies and methods, including an inventory of major building components, video imaging, checklist-style inspections, presentation graphics, and an interface to geographical information systems. Field experience with the BUILDER system will be used to evaluate its utility, accuracy, and cost effectiveness.

Another development in condition assessment practices, known as reliability centered maintenance (RCM), is "real-time" monitoring of building and equipment conditions based on concepts developed in the airline industry in the late 1960s and early 1970s. RCM has been used extensively in the aircraft, space, defense, and nuclear industries where functional failures can result in loss of life, have national security implications, and/or have extreme environmental impacts (NASA, 1996). The primary purpose of a traditional RCM program is to ensure safety at any cost; cost effectiveness is a secondary goal. A rigorous RCM analysis based on a detailed failure modes and effects analysis is used to determine appropriate maintenance for each identified failure mode and its consequences. RCM decisions are based on maintenance requirements derived from sound technical and economic justifications (NASA, 1996). An RCM program may not be the most practical or cost-effective approach for all facility maintenance programs.

The version of RCM being used by NASA is not, strictly speaking, a condition assessment process. RCM, "as defined and applied to facilities maintenance by NASA, is the integration of reactive maintenance (run-to-failure or breakdown maintenance), preventive (interval-based) maintenance, PT&I [predictive testing and inspection] (condition-based) and proactive maintenance. RCM applies these four techniques in combination where each is most appropriate based upon the consequences of equipment failure and its impact on organization, mission, safety, environment, and Life-Cycle Cost (LCC)" (NASA, 1996). The combination of techniques is intended to ensure the reliability of equipment and building components and minimize maintenance costs. The U.S. State Department, and possibly other federal agencies, are evaluating NASA's version of RCM as part of their comprehensive maintenance strategies.

ROLE OF TECHNOLOGY

Recent organizational downsizing has prompted facilities program managers to look to technology to provide facility-related data for decision making. Task 2 of the committee's charge was to investigate the role of technology in performing automated condition assessments. The committee reviewed (1) technologies that can support the collection and analysis of data for condition assessments and (2) technologies that could automate the condition assessment process itself.

Collection and Analysis of Data

With advancements in computer technology, pen-based data collection devices are now available that can guide inspectors through the inspection process, validate data in real time, and save time and eliminate errors as inspection data are uploaded to host systems.[6] Although these devices have a number of advantages, they also have some drawbacks: the average cost per unit is more than $2,000; they can be difficult to read in bright light; they can be cumbersome to use around mechanical and electrical equipment; they are susceptible to damage from falls and hard knocks; and they can be difficult to transport into tight spaces or up to rooftops.[7]

---

[6]Pen-based data collection devices have been used by DOE and DoD for both CASs and facility inventories.

[7]DoD inspectors used this system as part of the "Fence-to-Fence" CAS at 12 DoD Installations and the Facility Management System at Fort Riley, Kansas.

Another device, a bar code scanner, provides a relatively low-cost method of standardizing and uploading the data collection process but does not have the flexibility to adjust to unique situations or conditions. Bar code scanners typically have wand-type recording units that scan various condition codes carried by inspectors on cards or in booklets. The units are relatively durable in the field and are an inexpensive means to capture data as long as the inspection process is simple and does not require substantial data entry.

Recent developments include the integration of geographical information systems with video and inspection data through a mix of digital camera, automated data collection devices and Global Positioning System technologies to provide an integrated data process for assessing the condition of a system.[8]

Testing and Instrumentation

With testing and instrumentation, facilities program managers can monitor the overall condition of their key assets. The committee identified two types of testing that could be useful for federal facilities managers: real-time monitoring and systematic inspection.

NASA has an ongoing program of real-time monitoring, the RCM program, to manage the condition of physical assets, primarily machinery. Real-time monitoring, in the form of predictive testing and inspection, uses primarily nonintrusive testing techniques, visual inspection, and performance data to assess the condition of machinery. NASA now schedules maintenance only when it is warranted by the condition of the equipment rather than on a predetermined schedule. Continuing analysis allows NASA to plan and schedule maintenance or repairs to prevent catastrophic or functional failures and, at the same time, to optimize its resources (NASA, 1996). NASA also uses predictive technologies for monitoring the condition of critical building components, but not as extensively. The DOE now uses third parties for proactive system monitoring and instrumentation as part of their overall asset management programs. The committee did not identify any other federal agencies that use real-time monitoring to manage the condition of their physical assets.

Systematic inspections based on testing and instrumentation equipment as a means of monitoring the condition of individual systems are widely used by industry. Examples of technologies used in the private sector are listed in Table 3-1.

A number of nondestructive evaluation methods[9] for determining the condition of infrastructure, particularly of bridges and pavements, are available and in use in limited applications. "Nondestructive evaluation can be performed using a

---

[8]The University of Arkansas Mack-Blackwell Transportation Center and the North Carolina Department of Transportation use similar systems.

[9]Nondestructive evaluation involves the "use of tests to examine an object or material to detect imperfections, determine properties, or assess quality without changing its usefulness" (McGraw-Hill Encyclopedia of Science and Technology, 1997).

TABLE 3-1   Technologies Widely Used in the Private Sector

Type of Inspection	Time Interval
Thermographic analysis of electrical panels	2 to 5 year intervals
Eddy current analysis for chillers	5 to 8 year intervals
Roof moisture analysis	5 to 10 year intervals

wide range of technologies that include a simple visual survey to photographic methods to noncontact sensors" (Uddin and Najafi, 1997). These techniques are "most commonly used to evaluate the extent of damage discovered in a more traditional inspection, although some types of NDE [nondestructive evaluation] techniques can be used in place of visual inspection" (Sanford and McNeil, 1997). Some of these technologies include radar for evaluating structural integrity, delamination, layer delineation, voids, and moisture damage; and, infrared thermography for measuring temperature, detecting leaks, delamination, defective areas, and for stress mapping. "Acoustic and ultrasonic testing and laser technologies are being used for detecting cracking and defects, . . . faulting of transverse joints, and surface texture and displacement measurements" (Uddin and Najafi, 1997).

Data Management Systems

In recent years, computer-aided facility management systems (CAFM) and computerized maintenance management systems (CMMS), which are similar, but not necessarily congruent, have advanced. CAFMs and CMMSs are used in a wide range of applications, from facility and space planning to preventive maintenance and parts inventories to budgeting and project management. As these systems continue to mature, vendors are offering a wide range of hardware and software products and services, as well as tools to produce and manage information. Few standards have been established for these systems, however, and the technologies are changing rapidly. Federal program managers must first evaluate their agency's needs and objectives and reporting requirements before choosing a computerized system.

Decision Support

The increase in the availability of computer systems has provided one of the most attainable and useful technological advances for facilities program managers. Current decision-support technologies offer a wide array of facility analysis tools with the capability to project maintenance and repair requirements, determine life cycle costs, and prioritize needs.

The American Society for Testing and Materials (ASTM) has published information on a number of standard practices that can provide facilities program managers with automated analysis tools to determine benefit-to-cost and savings-to-investment ratios, internal rate of return, and net benefits for infrastructure investments (ASTM, 1992). In addition, a number of multiple-attribute decision-making methods have been developed and automated for use in evaluating facility maintenance and repair requirements based on multiple criteria.[10] These decision-support tools have been used for quite some time for the evaluation of weapons platforms, software systems, and research and development projects and could also provide facilities program managers with defensible methodologies for evaluating and prioritizing maintenance and repair requirements.

Intelligent Buildings

In 1988, the Building Research Board of the National Research Council undertook a study entitled Electronically Enhanced Office Buildings to provide guidance to project administrators, building owners, chief executive officers, and facility planners about incorporating modern electronic technology in new or renovated buildings (NRC, 1988). "Smart" or "intelligent" buildings were made possible by the convergent evolution of communication, control, and computer technologies that had the potential to create highly efficient, automated buildings that could be monitored for energy, environmental, and life safety parameters and which could then respond to either remote or preprogrammed control. These intelligent building systems are governed by straightforward feedback and control logic, which predated by many years the technologies that could implement them economically. However, the energy crisis of the 1970s and the resultant national focus on energy efficiency provided an economic motive for reducing building energy costs and ultimately led to the development and deployment of the necessary technologies. Today's intelligent buildings integrate sensor and monitoring devices, data transmission capability by means of telephone lines, fiber optic cable, or satellite uplinks, computers for data management and decision making, and microprocessor control devices. Most of the control decisions are preprogrammed and require little, if any, operator intervention. However, the concern that the realization of these buildings would fall far short of their potential has still not been fully addressed (Teicholz and Ikeda, 1995). Although the number of existing buildings that now offer one or more of these features has increased, the potential of these technologies has not been fully realized. In light of this, the committee attempted to determine if intelligent building technologies could be used for the monitoring, diagnosis, and management of potential maintenance problems, i.e., automated condition assessments.

---

[10]Current applications include Fort Riley's facility management decision support system and Dallas/Fort Worth International Airport's capital asset prioritization system.

Building Diagnostics

The concept of building diagnostics was defined in a previous NRC study (1985) as: a set of practices that are used to assess the current performance capability of a building and to predict its likely performance in the future.

The essential elements of a building diagnostics program are:

- knowledge of what to measure
- appropriate instruments and other measurement tools
- expertise to interpret the results
- a capability for predicting the future condition of the building

Ideally, a building diagnostics program would enable a facilities program manager to devise corrective procedures when the future condition is likely to be undesirable. Although all components of a building and its systems could be included in a building diagnostics program, as a practical matter, the process is particularly well suited to the components and systems with high consequences of failure.

Automated Condition Assessments

The idea of linking building diagnostics with the monitoring and control capabilities inherent in intelligent buildings is very appealing. Building diagnostics deal with the measurement and interpretation of data and the relationship of those data to expected building performance. Technology for monitoring, telemetry, processing, and control, as well as the capability to integrate these technologies, is well developed, if not fully mature, and is already deployed in many buildings. Automating the condition assessment process for high-consequence systems through intelligent building technologies would appear to be a small step. However, at the present time, such systems are not widely used (Smith, 1998).

Microprocessor technology has made it possible for manufacturers to include intelligent controllers on various types of mechanical equipment, such as heating, ventilating, and air conditioning systems, chillers, and boilers. Enabling these systems to communicate and share data with each other is a matter of systems integration. The capability for them to work together is referred to as interoperability. Integration can occur at both the equipment level (unified control of the various devices) and the information level (the ability to access and process data from intelligent equipment). One of the chief impediments to full interoperability has been the lack of a unified standard or protocol for communications between individual devices. The BACnet™ communication protocol for building operating systems is an attempt to overcome this impediment (ASHRAE, 1994) and could be an important step toward full and open systems integration.

Open systems integration is also a requirement for automated condition assessment systems. A simplified diagram of an automated condition assessment process is shown in Figure 3-1.

Beyond the "traditional" control of building mechanical systems, existing sensor and microprocessor technologies also have the potential to monitor and manage a range of environmental parameters that are difficult to inspect and measure during routine site visits and condition assessments. For example, interior building moisture levels behind walls and bulkheads may indicate that conditions are favorable to the growth of micro-organisms associated with Sick Building Syndrome and other adverse health effects. Sensor nets could be installed at the time of construction (or major renovations) to monitor and report on moisture levels and bring potential problem areas to the attention of facility managers long before they become serious. The same technology could also be used to monitor the integrity of the roof system.

Although there are numerous possible applications of currently available intelligent building technologies for monitoring and assessing buildings and their systems, proactive building diagnostic systems have not been widely deployed. The primary reason appears to be that economic paybacks to justify the initial cost of the systems themselves have not been well documented (Claar, 1998; Smith, 1998; Teicholz, 1998). Unlike process industries where automation and monitoring can be shown to have direct and quantifiable beneficial effects on the

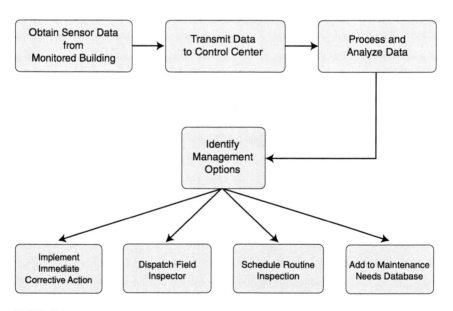

FIGURE 3-1   Logic and activity flow of an automated condition assessment process.

process, financial practices in the real estate industry do not always account for total life-cycle costs, which makes it difficult to justify the cost of automated systems (Petze, 1996). Unless a building owner is committed to reducing the life-cycle cost of a building and maintaining it at a high performance level, he is not likely to install these expensive systems.

The availability of a technology does not ensure its deployment. The technology must also be reliable and affordable and must clearly demonstrate benefits over existing methods. At this time, none of these three conditions can be met unequivocally for proactive building diagnostic systems. Although individual system components with acceptable reliability exist, sufficient experience with the components integrated into an automated condition assessment system does not. Affordability is a relative term, the absolute values of which can only be determined in application. Although the benefits of an automated system are assumed to include timely reporting (and avoidance) of problem conditions with associated long-term cost savings (as well as potential direct savings over existing condition assessment practices), data to support this assumption are not readily available.

Automating the condition assessment process offers the federal facilities program manager the potential for cost savings, improved building performance, and a means of coping with reduced staffing levels. The data to test these assumptions could be obtained either by designing and installing several systems in new federal buildings or by studying and evaluating buildings that already have them in the private sector. The federal government's responsibility for the long-term stewardship of buildings and facilities supports this kind of leadership role in deploying new building technologies and accepting higher first costs to reduce life-cycle costs.

Personnel Implications

Automated condition assessment processes will require different skills than are typically found in facility management and maintenance organizations. Intelligent buildings with building automation systems require facilities personnel to be familiar with a broad range of computer applications (graphics, databases, and spreadsheets) as well as hardware (personal computers and microprocessors). Personnel involved in automated condition assessments would be expected to have similar computer skills. Increasing the level of building sophistication will require staff to maintain and update the software systems and will require continuous training for operations and maintenance personnel. Although automated processes may require fewer personnel (or in turn be driven by them), the cost of highly trained, computer-literate facility technicians necessary to maintain these systems may offset any apparent savings attributable to downsizing. Therefore, decisions regarding the automation of monitoring and assessment processes should be based on a full life-cycle and building performance analysis.

ISSUES RELATED TO CONDITION ASSESSMENTS

The use of condition assessments by federal agencies is increasing. Attempts to catalog maintenance and repair deficiencies have evolved to include computerized programs with automated checklists linking condition assessment data to agency mission and improving facility management. NASA's condition assessment practices involve a sophisticated mix of regular inspections and predictive testing and instrumentation, as well as computerized programs. Federal agencies with condition assessment programs have generally developed them independently to meet their specific needs within resource constraints. Consequently, the level of sophistication varies widely.

Based on the information available to the committee, condition assessment programs, as currently practiced in federal agencies, are labor intensive, expensive to maintain, and time consuming. In theory, CASs provide excellent information as a basis for facilities management practices and maintenance and repair budget requests. In practice, the data are usually not provided in a time frame or format that is useful for cost-effective facilities management.

Cost of Data Collection

A database containing the numbers, ages, and sizes of buildings and similar "inventory" characteristics requires that data be gathered only once and updated when major changes occur, such as the acquisition of a new facility, the completion of a major addition, or the demolition of a building. The cost of establishing a database will depend on the number of facilities in the inventory. Once established, the maintenance of this type of database costs relatively little. In contrast, data related to the condition of buildings and their components must be gathered and updated on a regular basis to be useful. Condition-related data are usually gathered through inspections by trained personnel, often a team of several specialists. Facilities may be inspected on a two to five year cycle. Because of the personnel, equipment, data entry, and time involved in inspecting large inventories of buildings, the cost of data collection is high. For most agencies and organizations, "[t]radeoffs occur between the amount of data collected, the frequency at which it is collected, the quality of the data, and the cost of the entire process, including data entry and storage" (Sanford and McNeil, 1997).

Type and Volume of Data Collected

The federal agency condition assessment programs reviewed by the committee were designed to be comprehensive. Besides inspecting critical building components and systems, such as roofs, plumbing, electrical, and fire safety systems, agencies also collect data on more cosmetic deficiencies, such as broken door locks, and torn or worn carpeting. Having a team of inspectors identify both system and cosmetic deficiencies involves more time, resources, and data entry, and, therefore, higher costs than having them identify only critical component deficiencies. Including cosmetic deficiencies in a condition assessment program makes it more labor intensive and data intensive and, therefore, more expensive. More important, perhaps, the information collected becomes less timely, more difficult to analyze, and, consequently, less useful for ongoing facilities management.

Timeliness of Data

Current practices take so long to gather and analyze condition assessments that the information loses its value in the budget development process. Even one of the most sophisticated condition assessment programs, the Air Force's Commanders' Facility Assessment, has reported difficulties in keeping the data timely; in other agencies, the information is three to five years out of date. In a budget cycle that begins two years before the actual fiscal year, the information compiled through condition assessment programs loses much of its value for the development and justification of maintenance and repair budget requests.

REFERENCES

AME (Applied Management Engineering). 1991. Managing the Facilities Portfolio: A Practical Approach to Institutional Facility Renewal and Deferred Maintenance. Washington, D.C.: National Association of College and University Business Officers.

ASHRAE (American Society of Heating, Refrigeration, and Air Conditioning Engineers). 1994. BACnet - A Data Communication Protocol for Building Automation and Control Networks. Document SPC-135P-031. Atlanta, Ga.: American Society of Heating, Refrigeration, and Air Conditioning Engineers.

ASTM (American Society for Testing and Materials). 1992. Standards on Building Economics. Philadelphia, Pa:. American Society for Testing and Materials.

Claar, C. 1998. Personal communication from Charles Claar, Director of Research, International Facility Managers' Association, Houston, Texas to Richard Little, Director of Board on Infrastructure and the Constructed Environment, National Research Council, Washington, D.C. June 6, 1998.

Earl, R. W. 1997. The Condition Assessment Survey: A Case Study for Application to Public Sector Organizations. Pp. 277–286 in Infrastructure Condition Assessment: Art, Science, and Practice. Mitsuru Saito, ed. New York: American Society of Civil Engineers.

GAO (General Accounting Office). 1997. Defense Infrastructure: Demolition of Unneeded Buildings Can Help Avoid Operating Costs. Report to the Chair, Subcommittee on Military Installations and Facilities, Committee on National Security, U.S. House of Representatives. NSIAD-97-125. Washington, D.C.: Government Printing Office.

McGraw-Hill Encyclopedia of Science and Technology. 1997. "Nondestructive Testing," 8th ed., vol. 12, pp. 32–37. Chicago, Ill.: Lakeside Press.

NASA (National Aeronautics and Space Administration). 1996. Reliability Centered Maintenance Guide for Facilities and Collateral Equipment. Washington, D.C.: NASA.

NRC (National Research Council).1985. Building Diagnostics, A Conceptual Framework. Building Research Board, National Research Council. Washington, DC.: National Academy Press.

NRC. 1988. Electronically Enhanced Office Buildings. Building Research Board, National Research Council. Washington, DC.: National Academy Press.

NRC. 1990. Committing to the Cost of Ownership: Maintenance and Repair of Public Buildings. Building Research Board, National Research Council. Washington, D.C.: National Academy Press.

O'Hara, T. E., J.L. Kays, and J.V. Farr. 1997. Installation status report. Journal of Infrastructure Systems 3(2): 87–92.

Petze, J. D. 1996. Investing in Facility Automation - Improving Comfort, Air Quality, Building Management and the Bottom Line. Manchester, N.H.: Teletrol Systems, Inc.

Rugless, J. 1993. Condition assessment surveys. Facilities Engineering Journal 21(3): 11–13.

Sanford, K. and S. McNeil. 1997. Data Modeling for Improved Condition Assessment. Pp. 287–296 in Infrastructure Condition Assessment: Art, Science, and Practice. Mitsuru Saito, ed. New York: American Society of Civil Engineers.

Smith, T. 1998. Personal communication from Thomas Smith, Eastern Regional Manager, Teletrol Systems, Inc., Manchester, New Hampshire, to Richard Little, Director of Board on Infrastructure and the Constructed Environment, National Research Council, Washington, D.C. June 6, 1998.

Teicholz, E. and T. Ikeda. 1995. Facility Management Technology Lessons from the U.S. and Japan. Engineering and Management Press. Norcross, Ga.: Institute of Industrial Engineers.

Teicholz, E. 1998. Personal communication from Eric Teicholz, President, Graphic Systems, Inc., Cambridge, Massachusetts to Richard Little, Director of Board on Infrastructure and the Constructed Environment, National Research Council, Washington, D.C. June 6, 1998.

Uddin, W. and F. Najafi. 1997. Deterioration Mechanisms and Non-Destructive Evaluation for Infrastructure Life-Cycle Analysis. Pp. 524–533 in Infrastructure Condition Assessment: Art, Science, and Practice. Mitsuru Saito, ed. New York: American Society of Civil Engineers.

USACIR (U.S. Advisory Commission on Intergovernmental Relations). 1996. The Potential for Outcome-Oriented Performance Management to Improve Intergovernmental Delivery of Public Works Programs. SR-21. Washington, D.C.: U. S. Advisory Commission on Intergovernmental Relations.

Uzarski, D.R., and L.A. Burley. 1997. Assessing Building Condition by the Use of Condition Indexes. Pp. 365–374 in Infrastructure Condition Assessment: Art, Science, and Practice. Mitsuru Saito, ed. New York: American Society of Civil Engineers.

4

Strategic Framework

The ownership of real property entails an investment in the present and a commitment to the future. Ownership of facilities by the federal government, or any other entity, represents an obligation that requires not only money to carry out that ownership responsibly, but also vision, resolve, experience, and expertise to ensure that resources are allocated effectively to protect the value of that investment. Once facilities have been acquired, the long-term costs of maintaining them become the owner's responsibility.

Even though federal facilities represent investments of capital, they are generally not treated as capital investments from a management and accounting standpoint, as they would be in the private sector. In fact, there are fundamental differences between the objectives of the federal government and the objectives of the private sector and, consequently, in the ways they operate. Most importantly, the primary goal of business is to earn a profit whereas the goals of government are more complex and are often guided by issues of public health, safety, and welfare. Businesses invest in buildings and property to produce a return on invested capital, both through the rental income stream and through proceeds from the final disposition of the property. The decision to invest is based on considerations such as the cost of capital, depreciation over a fixed period of time, and tax strategies. The focus is very much on the owner's "bottom line," or financial return.

The Analytical Perspectives volume of the President's 1998 Budget states that "there is no single number or 'bottom line' for the Government comparable to the net worth of a business corporation" (OMB, 1997). Because of the absence of a financial bottom line, the government must use discretion when looking to the private sector for asset management strategies. Even though factors at what might be termed the strategic level (i.e., linking the need for a facility to agency mission and ensuring employee health and productivity) are similar, the financial motivations are quite different.

In the private sector, the owner of an investment property usually maintains a building in a condition that ensures a positive ROI throughout its economic life. The owner will take whatever actions are necessary to maximize the ROI including the amount realized when the property is sold. Owners of investment buildings will generally tailor the level of maintenance to the rent the tenants are willing to pay. Although this may be an excellent financial strategy, it does not ensure that the building will be well maintained.

An owner-occupied building in the private sector is more like a government-owned facility. Intangible factors, such as corporate image, quality of the physical environment, and employee satisfaction are considerations in the facility maintenance and repair function. Even in the case of a building that is nearing the end of its service (not economic) life, maintenance and repair are justified, as long as they support the achievement of other goals. However, even in this case, the government's financial considerations and incentives are different from those of the private sector. A government agency has many objectives in maintaining its facilities, but making a profit is not one of them. Nevertheless, the government is expected to sustain the taxpayers' investment in facilities.

Chapters 1-3 described issues and findings related to maintenance and repair of the federal facilities portfolio. To address these findings the committee attempted to develop a methodology and rationale that federal facilities program managers could use to systematically formulate and justify facility maintenance and repair budgets. However, the current state of practice, the lack of data, in general, and the lack of research results, in particular, precluded the development of a methodology per se. The committee instead developed a strategic framework of methods, principles, and strategies, which can serve as the basis for the development of a methodology for the systematic formulation of maintenance and repair budgets in the future.

The maintenance and repair of federal facilities is a complex issue and no single action or strategy will resolve all of the identified issues. Commitment at all levels of the federal government will be necessary over the long term to ensure that resources are optimized and the public's investment in the facilities portfolio is sustained. The overall goal of the strategic framework is to protect and enhance the functionality and quality of (and investment in) the federal facilities portfolio. The framework has two objectives: (1) to foster accountability for the stewardship of facilities at all levels of government; and, (2) to allocate resources strategically for maintenance and repairs. The strategies to achieve these objectives are outlined in Figure 4-1. A more detailed discussion follows.

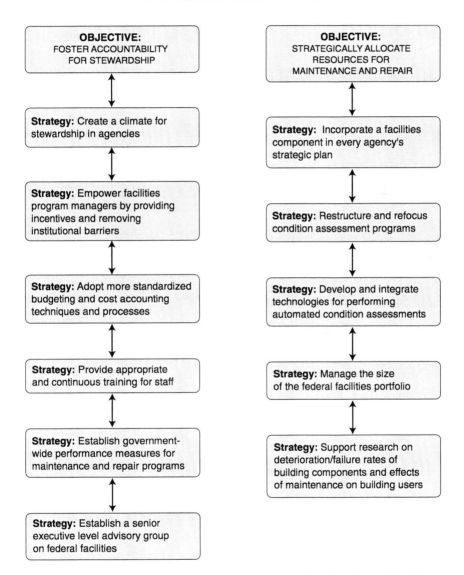

GOAL
Protect and Enhance the Functionality and Quality
of the Federal Facilities Portfolio

OBJECTIVE:
FOSTER ACCOUNTABILITY
FOR STEWARDSHIP

OBJECTIVE:
STRATEGICALLY ALLOCATE
RESOURCES FOR
MAINTENANCE AND REPAIR

Strategy: Create a climate for stewardship in agencies

Strategy: Incorporate a facilities component in every agency's strategic plan

Strategy: Empower facilities program managers by providing incentives and removing institutional barriers

Strategy: Restructure and refocus condition assessment programs

Strategy: Adopt more standardized budgeting and cost accounting techniques and processes

Strategy: Develop and integrate technologies for performing automated condition assessments

Strategy: Provide appropriate and continuous training for staff

Strategy: Manage the size of the federal facilities portfolio

Strategy: Establish government-wide performance measures for maintenance and repair programs

Strategy: Support research on deterioration/failure rates of building components and effects of maintenance on building users

Strategy: Establish a senior executive level advisory group on federal facilities

FIGURE 4-1   Strategic framework for the maintenance and repair of federal facilities.

FOSTERING ACCOUNTABILITY FOR THE
STEWARDSHIP OF FEDERAL FACILITIES

The responsible ownership of facilities by the federal government is an obligation that requires not only money, but also the vision, resolve, experience, and expertise to ensure that resources are allocated effectively to sustain the public's investment. The recognition and acceptance of this obligation is the essence of stewardship. Public officials and employees at all levels of the federal government are responsible for decisions that affect the stewardship of federal facilities:

- At the agency field level, facilities program managers develop budget requests for maintenance and repair funding, help determine which projects will be completed based on available funding, spend the funds allocated to them, and implement condition assessment programs.
- At the agency headquarters level, senior managers determine agency priorities among competing interests for operations and maintenance funding and for new construction; develop agency budget requests, including operations and maintenance requests; defend budget requests to OMB and Congress; allocate appropriated funds back to the field level; and make adjustments between operating and maintenance and repair funds during the fiscal year.
- Oversight agencies, such as OMB, review and revise agency budget requests, monitor the spending of appropriations, and rescind funds not obligated during the fiscal year.
- Congress reviews agency budget requests, appropriates operating funds, approves and funds the acquisition of new facilities, and legislates requirements for public health, safety, and welfare.

Because all levels of government share the decision-making responsibility for the stewardship of federal facilities, no single entity can be held responsible or accountable for the results. The lack of accountability for maintenance and repair issues has created a climate in which the long-term care of the facilities investment is regularly sacrificed to current programs. Effective stewardship of the federal facilities portfolio requires that public officials and employees accept their long-term as well as short-term obligation; recognize the long-term consequences of short-term actions; and make a commitment not to compromise the long-term viability of the federal facilities portfolio for more immediate concerns.

Greater accountability requires that public officials and employees be held responsible for the consequences of their actions. It also requires that they be given the appropriate tools, funding, and authority to carry out the responsibilities for which they are to be held accountable. The strategies set forth on the following pages are intended, in part, to create a climate for stewardship that will foster a shared assumption that public officials and employees at all levels of government must do their best not to sacrifice the functionality and quality of federal buildings to short-term interests.

Strategy: Create a Climate for Effective Stewardship in Federal Agencies

The four most important elements in creating a climate that encourages effective stewardship of facilities in federal agencies are:

- leadership by agency senior managers
- the establishment and implementation of a stewardship ethic by facilities program managers and staff as their basic business strategy
- senior managers and program managers who create or seek incentives for successful and innovative facility management programs
- agency strategic plans that give suitable weight to effective facilities management

There is no substitute for an agency director and senior managers who recognize the importance of facilities maintenance and operations and who will fight for the resources needed to maintain facilities effectively. A chief executive willing to "make waves" both within and outside an agency on behalf of facilities immediately frees internal staff from two burdens: first, they no longer have to find the "right" arguments to persuade the executive to take up their cause, and second, they can be confident that the need for facilities maintenance is being addressed during budget formulation and negotiation. These anxieties are a powerful disincentive to facilities program managers to spend time or intellectual creativity advocating measures or funding necessary to protect the facilities investment.

A chief executive who does not provide leadership in this area may spend much of his/her tenure combating indifference and engaging in crisis management and damage control on facilities-related issues. An effective agency director will support a substantive role for facilities program managers in the agency's strategic planning process, in the development of budget requests for facilities maintenance and repairs, and in finding innovative, cost-effective ways to implement maintenance and repair activities.

Facilities program managers should place their leadership in a context of innovation and stewardship. They should provide good fiscal analyses and a convincing rationale that chief executives and senior managers can use to advocate long-term facilities interests against short-term operations interests in the budget arena. The basic rationale for any government facility is that it is a long-term investment that supports mission-related activities. Facilities program managers should consider it their job to defend long-term interests against more parochial interests. Facilities program managers who do not take stewardship seriously will not be in a position to accomplish much that is useful.

In order for facilities program managers to practice effective stewardship, they must have the authority, "tools," and incentives to implement cost-effective programs for maintenance and repair. If they are given the necessary tools, are empowered and encouraged to use modern maintenance and repair management methods, and are rewarded for innovations, initiatives, and risk taking, then they can also be held accountable for maintenance and repair performance. They should be held accountable for knowing and properly reporting on the condition of federal facilities; for developing appropriate maintenance and repair budget requests; and for the efficient use of appropriated funds.

Strategy: Empower Facilities Program Managers and Remove Institutional Barriers

In today's environment, federal facilities program managers are faced with extending the useful life of aging facilities; altering or retrofitting facilities to consolidate space or accommodate new functions and technologies; meeting evolving standards for safety, environmental quality, and accessibility; maintaining or disposing of excess facilities; and finding innovative ways and technologies to maximize limited resources. In order to meet these challenges, they must be provided with adequate resources, appropriate authority, training, "tools," and incentives.

Motivating personnel to change long-established processes is a difficult undertaking. Employees need incentives to work toward improvement or to take risks that could (or could not) result in cost savings. The federal budget process for maintenance and repair activities does not offer incentives for improvement or for lower costs. In fact, it has disincentives. Agencies may be penalized if they do not obligate their entire budgets within a given fiscal year even if carrying over unobligated funds into the next fiscal year can be shown to be cost effective.

There are many opportunities for lowering facilities costs. Public agencies have been conducting a wide variety of experiments in improving public sector operations and modernizing public agency management, including outsourcing or privatizing facility-related functions or securing services through performance-based contracts. New opportunities, however, carry risks. There is always the possibility of failure in trying something new. Encouraging facilities program managers to be innovative and to take risks means offering them rewards commensurate with the risks. It also requires that some level of experimentation and failure be tolerated, although the causal factors leading to failures should be researched and documented so that mistakes are not repeated.

One potential reward for facilities program managers would be to allow them to take savings from one area of operations and maintenance and apply it to another. For example, if energy costs are lowered through cost-effective management, the savings could be used to finance other facilities-related improvements. To foster accountability, as well as recognize the achievement, the savings should be documented and should appear in the budget in a way that identifies how the savings will be used. Recognition awards for facilities maintenance and repair programs that achieve high levels of performance over a sustained period of time could also be incentives for facilities program managers to seek continuous improvement in maintenance and repair practices.

Facilities program managers may also improve the efficiency of operations by consolidating functions or leasing out space. For instance, "by consolidating 14 laundry facilities over a 3-year period, the VA [Veterans Affairs] expects to achieve a one-time equipment and renovation savings of about $38 million as well as recurring savings of about $600,000 per year from operational efficiencies" (GAO, 1996). The Army and other agencies have facility use contracts "at nine inactive ammunition plants and at all of them the ARMS [Armament Retooling and Manufacturing Support Act of 1992] initiative has produced revenue to offset all or part of the maintenance costs" (GAO, 1997b).

Agencies might also be allowed to sell excess properties and retain the resulting funds for repair and maintenance of other, mission-related facilities, where appropriate. Given the pressures on agency senior managers and executives to focus on short-term operations instead of long-term issues of stewardship, however, a government-wide policy allowing agencies to retain funds from real estate sales in all circumstances could lead to unintended consequences. Agencies trying to raise funds for operating programs could be tempted to hold "fire" sales of properties that may be needed to meet future mission requirements. Program managers and agencies that can document and justify the sale of a property that is not needed to meet long-term mission requirements, however, should be able to retain the savings and apply them to other identified maintenance and repair needs, on an agency by agency basis. To ensure accountability, the accounting system should clearly track how proceeds are spent and identify the results.

No one can guarantee that empowering managers and removing institutional barriers will not result in some failures or abuses of the system. Some level of failure must be tolerated to encourage innovation. Agencies and managers should be encouraged to set up pilot programs to test new tools, technologies, and strategies for cost-effective maintenance and repair. The expectations, objectives, costs, benefits, and outcomes of these test programs should be shared with other agencies so that successes can be duplicated and failures avoided.

Revolving funds are a "tool" which offer several potential advantages for maintenance and repair. "Revolving fund activities operate with no, or very little, direct appropriated funds. They are instead financed on a reimbursable basis from appropriated funds. These funds are appropriated to customers of these activities . . . who, in turn, purchase goods and services from the revolving fund activity much like any private business" (NPR, 1993). Revolving funds would also enable agencies to accumulate the resources to make capital acquisitions over time; to establish more consistent revenue streams; and to overcome the disincentives created by the end of fiscal year "spend it or lose it" budget process. They would allow facilities program managers to consider the full costs and benefits of proposed actions and to make up-front investments that could have long-term paybacks in operating efficiencies. Revolving funds could be used to reduce the backlog of maintenance and repairs, fund major repairs and replacements, or pay for unfunded legislative requirements.

Revolving funds require good financial management and oversight (GAO, 1997a). A system that holds managers accountable for the use and solvency of revolving funds should be established. The U.S. Navy Public Works Center Revolving Fund is a model already in use.

Strategy: Create Accountability through More Standardized Budgeting and Cost Accounting for Facilities Management and Maintenance

One of the findings of this study is that it is difficult, if not impossible, to determine how much money individual federal agencies and the federal government as a whole appropriate and spend for the maintenance and repair of federal facilities. The difficulty can be attributed to variations across agencies in budgeting procedures, definitions, and accounting structures, the structure of operations and maintenance budgets, the moving of funds between operations and maintenance activities, and the lack of tracking of maintenance and repair expenditures.

A single, standardized, government-wide system for developing maintenance and repair budgets and accounting for maintenance and repair expenditures may not be possible, or even desirable. A single system might not accommodate the variations among agency missions and programs, and, more important, the benefits of a single system might not outweigh the costs (e.g., time and salaries) involved in designing and implementing it. However, greater standardization in budgeting, cost accounting, defining activities, and calculating facilities-related terms across government agencies would increase accountability for the stewardship of federal facilities.

The authoring committee of this report developed an illustrative template for facilities-related activities (see Figure 4-2) that can be used as a meaningful first step in the development of a more standardized budgeting and cost accounting structure. The template could be used as a "tool" by facilities program managers to formulate and justify maintenance and repair budget requests in the context of a facilities management program and to track expenditures. Senior managers, oversight agencies, and decision makers could use the information in the template to gain a better understanding of an agency's overall facilities-related activities and to evaluate the potential impact of their decisions. If most or all government agencies used a similar template for formulating budget requests and tracking allocations and expenditures, in conjunction with their existing systems, comparisons of budget expenditures could be made and standardized measures developed. These measures could then be used to develop benchmarks and to identify best practices for facilities portfolio management and maintenance.

The template is structured to show all of the major costs of ownership of a facilities inventory, i.e., routine maintenance, repairs, and replacements, facilities-related operations, alterations and capital improvements, legislatively mandated activities, new construction and total renovation activities, and demolition, as well as their interrelationships. The purpose is to foster clearer accountability for

Facilities Management-Related Activities	Included in 2-4% Benchmark	Funding Category and Comments
A. Routine Maintenance, Repairs, and Replacements • recurring, annual maintenance and repairs including maintenance of structures and utility systems, (including repairs under a given $ limit, e.g., $150,000 to $500,000 exclusive of furniture and office equipment) • roofing, chiller/boiler replacement, electrical/lighting, etc. • preventive maintenance • preservation/cyclical maintenance • deferred maintenance backlog • service calls	Yes	Annual operating budget
B. Facilities-Related Operations • custodial work (i.e., services and cleaning) • utilities (electric, gas, etc./plant operations) • snow removal • waste collection and removal • pest control • security services • grounds care • parking • fire protection services	No	Annual operating budget
C. Alterations and Capital Improvements • major alterations to subsystems, (e.g., enclosure, interior, mechanical, electrical expansion) that change the capacity or extend the service life of a facility • minor alterations (individual project limit to be determined by agency $50,000 to $1 million)	No	Various funding sources, including no year, project-based allocations such as revolving funds, carryover of unobligated funds, funding resulting from cost savings or cost avoidance strategies
D. Legislatively Mandated Activities • improvements for accessibility, hazardous materials removal, etc.	No	Various sources of funding
E. New Construction and Total Renovation Activities	No	Project-based allocations separate from operations and maintenance budget. Should include a life-cycle cost analysis prior to funding
F. Demolition Activities	No	Various sources of funding

FIGURE 4-2 Illustrative template to reflect the total costs of facilities ownership.

the allocation, expenditure, and tracking of funds for federal facilities at all levels of government and to provide greater visibility for the total costs of facilities ownership and its major elements. The template is a significant departure from current budget practices, which account for new facilities acquisition, major repairs and replacements, capital investments, and operations and maintenance separately but which blur the lines between various maintenance and operations functions. However, the template can be useful within current budget practices. In this template, the full costs of ownership are visible. The source of facilities-related funding is less significant than the total amount of funding requested, allocated, and obligated for an entire facilities management program.

Six significant activities are included: routine maintenance, repairs, and replacements; facilities-related operations, which includes activities such as snow removal and custodial work; alterations and capital improvements, which includes major capital expenses that recur on a 5, 10 or 20 year basis, and minor alterations; legislatively mandated activities such as asbestos removal; new construction and total renovation projects; and demolition. Effective use of the template concept can also give visibility to the value of and promote a more detailed chart of accounts and some standardization of definitions and calculations to facilitate comparisons across agencies. Further development of the template concept should be the responsibility of an advisory body of senior level federal managers, other public sector managers, nonprofit and private sector representatives (described later in this report). The elements that should be included under each activity and potential sources of funding are described below.

Routine Maintenance, Repairs, and Replacements

This activity includes maintenance and repairs that recur on an annual basis. Recurring maintenance and repair includes preventive maintenance (planned, scheduled periodic inspections, adjustments, cleaning, lubrication, parts replacement, and minor repairs of equipment and systems); preservation or cyclical maintenance, such as flood-coating of roofs, service calls for unscheduled or unanticipated maintenance, and routine replacements. The need to replace an item or system may arise from obsolescence, cumulative effects of wear and tear, premature service failure, or destruction by fire or other hazards (NRC, 1990). Replacements do not significantly increase the capacity of the item involved; such work is considered routine maintenance and repair if it is required for the continued operation of a facility (FFC, 1996).

In most cases, the annual operations and maintenance appropriation is the source of funding for all these activities. The total funds allocated by an agency for these activities are part of the total maintenance and repair funds that should be measured against the benchmark of 2 to 4 percent of the aggregate current replacement value of the agency's facilities inventory.

Facilities-Related Operations

This activity includes items associated with the routine operations of facilities but that do not include a significant amount of maintenance and repair work. For effective tracking of maintenance and repair costs, facilities-related operations should be accounted for separately. Funding for operations should not be counted against the 2 to 4 percent benchmark, because including such activities will not give a true picture of maintenance and repair expenditures and will skew any results that may be used for benchmarking purposes.

Alterations and Capital Improvements

These activities and their costs are incurred once, infrequently, or irregularly during the service life of a facility. Capital improvements include major alterations to structural or mechanical systems that change the capacity or extend the service life of a facility. Minor alterations may include reconfigurations of space and similar activities that do not include significant maintenance and repair work. Funding for these activities may be project-based, rather than from the annual operations budget, because they do not occur on an annual basis and can involve significant costs.

Legislatively Mandated Activities

This activity includes facilities-related projects undertaken in response to legislative requirements. These projects include retrofitting facilities for accessibility and removing hazardous materials, such as asbestos and underground storage tanks. These are typically one-time improvements with costs comparable to the costs of other capital projects. The usual source of funding is the annual operations and maintenance appropriation. Funding for any of the activities could also come from the carryover of unobligated funds, where justified, funds generated internally by cost-saving strategies, revolving funds, or other innovative measures. The committee recommends that these activities be accounted for separately to provide facilities program managers and public officials with a better understanding of the total costs of unfunded legislative mandates and their impact on operations and maintenance budgets.

New Construction or Total Renovation Activities

New construction or total renovation activities above an established dollar threshold (which varies by agency) are currently separate line items in the federal budget and are not funded from the operations and maintenance appropriation. These projects are important elements of an agency's total facilities management program, however, and should be accounted for in the template in order to evaluate their long-term impact on maintenance and repair resources for existing facilities and to provide an accurate picture of the total costs of facilities ownership. Tracking the number and square footage of new or totally renovated facilities will also help agencies to better track the size of their facilities portfolios.

Demolition Activities

Demolition projects are usually funded from the annual operations and maintenance appropriation, although they can be funded as separate, "earmarked" authorizations. Demolition is an important component of a facilities management program and the total costs of ownership. Accounting for demolition projects separately will allow program managers to predict where future cost savings may occur. The total number and square footage of facilities demolished should be tracked to monitor the size of the facilities portfolio.

Strategy: Provide Training in Facilities-Management Principles for Staff That Develops or Reviews Maintenance and Repair Budgets

As an inherently governmental function, preparation and review of agency budget requests must be done by federal personnel. Because of administration and congressional policies to reduce agency staffing, experienced personnel have been leaving the government in record numbers. Agency managers are concerned that institutional knowledge and technical expertise are being lost in the process. The loss of technical expertise has significant implications for the maintenance and repair function because a lack of sensitivity to the total costs of facilities ownership is one reason for the long-term underfunding of maintenance and repair programs. In the collective experience of the committee members, budget analysts do not necessarily have technical expertise in the budget area being analyzed. A firm grounding in the principles of facilities management and an understanding of the relationship between adequate and timely maintenance and repair to the total costs of facilities ownership is a critical skill. Facilities constitute a significant public investment that needs to be protected through cost-effective practices based on informed decisions. The people who prepare or review facilities management budgets should be trained in the principles of facilities management and related topics, and this training should be updated continuously.

Strategy: Create Government-Wide Performance Measures for Maintenance and Repair Activities

Performance measures are critical elements of a comprehensive management system for facility maintenance and repair. Determining how well the maintenance function is being performed or how effectively maintenance funds are being spent, requires well defined measures. Performance measures to evaluate the effectiveness of facilities maintenance and repair programs could promote greater accountability for facilities stewardship throughout the federal government. Unlike budgeting practices that have evolved over time and are ingrained into agency procedures and cultures, performance measurements are not well established. However, because performance measures are required by the Government Performance and Results Act of 1993, agencies have been attempting to develop them on their own. Because performance measures are not yet ingrained into agency processes, an opportunity exists to develop facilities management-related performance measures that can be used throughout the government.

Consistent measures would allow for comparisons within and across agencies of the effectiveness of facilities management programs, including maintenance activities. For example, a potential indicator of physical condition might be the number or percentage of facilities rated as not meeting life, health, or safety codes and standards. Uniform measures are especially important for organizations with decentralized functions, in which individual centers and installations have discretion over how funds are spent. Comparisons of performance within and across agencies would help to identify "best practices" and would contribute to overall improvements in the management and maintenance of facilities.

A study by the National Research Council, Measuring and Improving Infrastructure Performance, found that "no adequate, single measure of performance has been identified, nor should there be an expectation that one will emerge. Infrastructure systems are built and operated to meet basic social needs, but those needs are varied and complex" (NRC, 1995). Therefore, the measures used to evaluate facilities and infrastructure performance should vary. The report goes on to state that "Infrastructure performance is the degree to which infrastructure provides the services that the community expects of that infrastructure, and communities may choose to measure performance in terms of specific indicators reflecting their own objectives" (NRC, 1995).

The report concluded that these indicators generally fall into three broad categories, measuring performance as a function of *effectiveness, reliability*, and *cost*. "Infrastructure that reliably meets or exceeds broad community expectations, at an acceptably low cost is performing well." Although this was a study of infrastructure systems at the community level, the principle that the performance of facilities maintenance and management functions can, and should be, measured by the condition of the facilities inventory as measured against cost and effect on agency mission, is also applicable to the maintenance and repair of federal facilities.

Although it may appear that mission readiness and cost alone are insufficient to judge the performance of the maintenance and repair function, if the measures are broad enough, they will capture all relevant aspects of a facility's condition. For example, the health, physical comfort, and morale of the employees who occupy government buildings are also factors that could, with more research, be directly related to mission readiness. Minimizing costs, which is often the controlling factor in a unidimensional, budget-driven decision system, may undervalue these factors in determining cost-effectiveness. How these factors are weighed in their application will be determined in large part by the nature of the facility, the severity of the impact, and if feasible, how performance is measured. Maintaining the physical appearance and user accessibility for national landmarks, such as the U.S. Capitol, is more important than for a purely administrative facility. Military installations with a combat readiness mission have different priorities than administrative facilities. Not all facilities on an installation contribute equally to combat readiness, so condition priorities will also vary at the installation level.

Even though the stakeholders involved with the maintenance and repair of federal facilities differ significantly from the stakeholders described in Measuring and Improving Infrastructure Performance, the same principles apply. In both cases, the stakeholders, i.e., the agency directors, facilities program managers, building operators, building occupants, and other customers of the facility, must judge whether the maintenance and repair program contributes to mission achievement for the agency. Performance measures for facilities maintenance and repair can be developed from the principles recommended in Measuring and Improving Infrastructure Performance. Once a preliminary set of measures has been developed, they can be tested for effectiveness and applicability by evaluating the maintenance and repair functions for a few agencies before being applied government-wide.

Strategy: Establish a Senior-Level Advisory Group on Federal Facilities Issues

Accountability for the stewardship of federal facilities at the highest levels of government is at least as important as accountability at the agency and field office level. Senior leadership has the responsibility to encourage cost-effective management of the federal facilities portfolio to protect the public's investment. The authoring committee recommends that an executive level, federal facilities advisory group be appointed to provide policy direction and set priorities for the effective management and maintenance of the facilities portfolio. This group should include senior level federal managers from civilian and military agencies, other public sector managers, and representatives of nonprofit organizations and private sector corporations. Models for this type of policy advisory group include the Federal Facilities Policy Group, which was convened by the director of the OMB and the chair of the Council on Environmental Quality to review the status and future course of environmental response and restoration of federal facilities. A report by the Federal Facilities Policy Group "identified areas of management and regulatory reform essential to protect public health and restore the environment as well as assure effective, efficient use of resources as the effort to clean up federal facilities proceeds" (COEQ and OMB, 1995).[1]

An advisory group of senior officials from DoD, DOE, GSA, NASA, other federal agencies responsible for managing facilities portfolios, OMB, GAO, the National Science and Technology Council, and other appropriate agencies and organizations should be appointed to focus on the policy issues related to maintaining and enhancing the functionality and quality of federal facilities. This group should also include representatives of state and local governments, nonprofit organizations, and private sector corporations with facilities-related responsibilities to provide a broad perspective on facilities management. An executive level advisory group will give the issue of federal facilities maintenance, repair, and stewardship greater visibility. Initially, this effort may require the investment of more staff time and resources, but, in the long term, it should result in savings of both time and resources through greater cooperation and sharing of facilities management knowledge.

Initial focus areas for the advisory group could include:

- identifying the entities responsible for facility stewardship at all levels of government, outlining their responsibilities, defining reporting requirements, and setting standards for accountability
- providing guidelines for developing government-wide performance measures for evaluating the effectiveness of facilities management and maintenance programs
- establishing a process to identify best practices for facilities maintenance and repair
- developing a standardized, annotated chart of accounts for agencies to use in developing maintenance and repair budget requests and for tracking allocations and expenditures (A standardized chart of accounts, which could be based on the template described above, would facilitate interagency exchanges of information, the roll-up of data into broad classes for comparative analyses, and benchmarking for facilities management and maintenance and repair, inside and outside the federal government. A process for reviewing the quality of the data could also be developed.)
- identifying ways to eliminate institutional disincentives to effective facilities management and identifying potential incentives for innovative, cost-effective facilities maintenance and repair
- developing standardized methodologies for analyzing the life-cycle costs of facilities

---

[1]The Federal Facilities Policy Group included policy officials from the U.S. Departments of Agriculture, Defense, Energy, Health and Human Services, Interior, and Justice, the Environmental Protection Agency, NASA, and the U.S. Army Corps of Engineers. Several White House offices, including the President's Council of Economic Advisors, the Domestic Policy Council, the National Economic Council, and the Office of Science and Technology Policy, also participated.

- developing a decision model for reviewing agency portfolios, determining the value of facilities that are not considered mission-critical, and determining their appropriate disposition
- developing a government-wide database on excess facilities and a process for considering how excess facilities could be used and streamlining the process for the disposition of excess facilities
- establishing a process for comparing unit costs for common maintenance elements and for comparing service quality elements, such as breakdown/ problem incident rates

STRATEGIC ALLOCATION OF RESOURCES

Strategy: Incorporate a Facilities Component in Every Agency's Strategic Plan

The Government Performance and Results Act of 1993 was enacted "to provide for the establishment of strategic planning and performance measurement in the Federal Government and for other purposes." The Act is intended to accomplish the following goals:

(1) improve the confidence of the American people in the capability of the federal government by systematically holding federal agencies accountable for achieving program results
(2) initiate program performance reform through a series of pilot projects for setting program goals, measuring program performance against those goals, and reporting publicly on their progress
(3) improve federal program effectiveness and public accountability by promoting a new focus on results, service quality, and customer satisfaction
(4) help federal managers improve service delivery by requiring that they plan for meeting program objectives and by providing them with information about program results and service quality
(5) improve congressional decision making by providing more objective information on achieving statutory objectives and on the relative effectiveness and efficiency of federal programs and spending
(6) improve the internal management of the federal government

By September 30, 1997, the head of every federal agency was required to submit to Congress and OMB a strategic plan for his or her agency's program activities. The strategic plans were required to include comprehensive mission statements covering the major functions and operations of the agency, outcome-related goals and objectives, a description of how the goals and objectives were to be achieved, the resources needed to achieve them, and performance measures, among other items.

Congress and federal agencies have been discussing and reevaluating agencies' missions, functions, and responsibilities. Nevertheless, the relationship of constructed facilities to agencies' missions has largely been overlooked. Agency programs are, in fact, the driving force behind the construction and acquisition of federal facilities, which play supporting, but critical, roles. It is difficult to imagine for example, how the Smithsonian Institution could fulfill its mission without its museums, how the National Institutes of Health and the National Institute of Standards and Technology could fulfill their missions without research laboratories, or the Bureau of Prisons without prisons.

Agencies and Congress do consider how facilities help implement an agency's mission when budget requests for constructing or acquiring new facilities are reviewed. Once the facilities have been built or acquired, however, their relationship to the agency's mission is taken for granted, even though deteriorating facilities can seriously impair the fulfillment of an agency's mission. A recent GAO report found that "At the Naval Station in Norfolk, about half the piers are 50 years old and too narrow to accommodate today's larger ships. Many piers were in poor condition and, according to Navy officials, limited the Navy's ability to berth ships in transit and support its dock side requirements, such as loading supplies" (GAO, 1997c). Deteriorating facilities can also affect an agency's ability to recruit qualified employees, the productivity of current employees, and the efficient operation of the agency, all of which are related to effective implementation of the agency's mission.

As agencies reevaluate and, in some cases, redefine their missions, the relationship between mission and facilities should be made explicit. Federal agencies require adequately maintained facilities to accomplish their missions. For NASA to conduct space exploration, the agency needs well maintained launch facilities, research laboratories, and command control centers. Similarly, the Navy needs well maintained docking facilities for the fleet, and the federal judicial system needs well maintained courthouses to hold trials and store legal documents. All federal agencies need well maintained administrative buildings to deliver services to the public and provide healthy and productive environments for their employees.

However, not all facilities owned by federal agencies are critical to mission delivery. Some agencies already recognize that they own excess and/or under-utilized facilities. In reevaluating their missions, they may find that even more facilities will become "excess," or less critical, to support their current and future programs. Once an agency has determined which facilities are mission-critical, it can prioritize its needs for maintenance and repair and direct available funding to the facilities that are most closely linked to agency mission. Linking facilities to mission will enable agencies to link maintenance and repair budget requests and allocations to the long-term strategic planning of federal agencies and the government as a whole.

Strategy: Restructure and Reprioritize Condition Assessments

Federal agencies, such DoD, DOE, and others, conduct comprehensive condition assessments of all of their facilities every two to five years. Despite advances in technology, the type and volume of data collected and the number of facilities inspected makes condition assessments an expensive, time consuming, and labor intensive process. Because of the long time required to collect and analyze these data, their value for ongoing facilities management or developing maintenance and repair budget requests is limited. The more detailed the inspection, the higher the cost.

Federal agencies should determine if the benefits of comprehensive condition assessments are commensurate with the costs. Every agency should review the objectives of its condition assessment survey process in relation to its mission. Once the agency determines the data that are necessary to meet mission objectives and the costs of gathering those data, it can work "backwards" to determine the required level of detail (see Figure 4-3). Focusing condition assessments on information that supports facilities management and decision making would provide the best return on the investment of staff time and resources.

Agencies should consider restructuring their programs to focus first on facilities that are mission-critical; on life, health, and safety issues; and on the building components that are critical to a building's performance. The latter include the building envelope and system components, such as roofs, plumbing, and electrical systems. Data gathering should focus on compiling information that is critical to building performance, to building users' safety and health, and to informed decision making by the agency senior managers who review and formulate operations and maintenance budgets. Peeling paint, damaged carpets, and other nonstructural problems should not be included in condition assessments unless they indicate structural deficiencies (e.g., paint is peeling because the roof leaks), or they constitute a significant risk to the health, safety, or welfare of the building's

Typical Engineering Decision Process

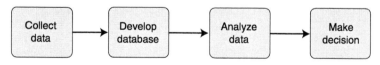

Reverse Engineering Decision Process

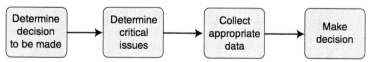

FIGURE 4-3   Engineering decision processes.

users. Condition assessments should concentrate on critical elements that affect the ability of an agency to operate effectively rather than simply cataloguing all problems, structural and cosmetic.

Condition assessment programs are sometimes structured to assess every facility in an agency's inventory once every two to five years. Available resources could be used more effectively if condition assessment programs were re-prioritized to inspect facilities that were identified as mission-critical by the agency first. These facilities might also require more frequent inspections than buildings that are not mission critical. Other facilities should continue to be inspected but at a level commensurate with their relationship to the agency's mission and to meeting life, health, and safety standards.

Because current condition assessments are time consuming and comprehensive, the information gathered in these inspections loses its value in developing agency maintenance and repair budget requests because data are not available in a suitable form or time frame. The condition assessment process should be streamlined and refocused. By linking condition assessments to mission-critical facilities, focusing on life, safety, and health standards, as well as critical building system components, they can be integrated into strategic planning and budgeting processes.

In summary, the purpose of restructuring and reprioritizing condition assessment programs is threefold: to make better use of available resources; to focus on deficiencies that can shorten a facility's service life, effect implementation of the mission, or pose significant risks to occupants' health or safety; and to collect meaningful, timely information that can be used in the budget process.

Existing and emerging technologies offer promise for automating the condition assessment process although many of these technologies are not yet widely deployed. The federal government's responsibility for the long-term stewardship of buildings and facilities supports taking a leadership position in deploying new technologies and accepting higher first costs to reduce life-cycle costs. The federal government and private industry should work together to develop and integrate technologies for performing automated facility condition assessments and eliminate barriers to their deployment.

Strategy: Manage the Size of the Federal Facilities Portfolio

As a result of decisions made over many decades, some agencies in the federal government now own more facilities than they need to perform their missions effectively. Responsible stewardship of federal facilities requires that the size of an agency's portfolio be commensurate with a level that supports the long-term mission of the government as a whole. The committee recommends two strategic approaches to the effective management of the size of the facilities portfolio: limiting the construction and acquisition of new facilities; and reducing the number of facilities owned and maintained by the federal government.

Limiting the Construction and Acquisition of New Facilities

Over the long term, the federal government will have to acquire new facilities to meet changing missions, circumstances, and technologies. New embassies will be needed as a result of changes in international politics and boundaries; technologically advanced buildings and facilities will be required for state-of-the-art research or to support new weapons systems; and other new buildings will be necessary to support as yet unidentified missions. Every new facility, however, entails a responsibility for the government to operate and maintain it for 30 years or more. At a time when some federal agencies already own excess facilities, every effort should be made not to increase the size of the federal facilities portfolio unnecessarily. New facilities should only be acquired after a rigorous analysis clearly demonstrates that a new facility is the best way to meet mission requirements and the most cost-effective course of action over the life-cycle of the facility.

Before acquiring a new facility, agencies should demonstrate that mission requirements cannot be met effectively through the use of existing facilities, either in their current or modified configurations. Agencies should consider using or adapting not only the facilities they own, but also excess or underutilized facilities owned by other agencies. Existing facilities embody already allocated resources, and maximizing their use can be cost effective and can help manage the size of the federal facilities portfolio.

If an agency demonstrates that existing facilities cannot meet mission requirements in a resource-effective manner, the agency should consider whether leasing a new facility would be more cost effective than acquiring one. In some cases, particularly if a facility will be needed for a relatively short time (10 to 15 years), leasing may be more cost effective over the life-cycle of the building. Once a leased facility is no longer needed, the federal government would not be responsible for its operation and maintenance or for its disposition. In contrast, the government continues to be responsible for the maintenance and disposition of owned facilities, even if they are no longer needed.

When considering the acquisition of new facilities, public officials have traditionally focused on the "first costs" of design and construction, which represent only 5 to 10 percent of the total costs of ownership of a facility. The federal budget process is structured to reinforce the emphasis on first costs. Understanding the full costs of acquiring and operating a facility over the 30 or more years of its service life, requires a rigorous life-cycle cost analysis, which has been defined as "the present value of all anticipated costs to be incurred during a facility's economic life; the sum total of direct, indirect, recurring, non-recurring and other related costs incurred or estimated to be incurred in the design, development, production, operation, maintenance, support, and final disposition of a major system over its anticipated useful life span" (NRC, 1993). As part of the analysis, the projected asset value the facility should be established in terms of the intended length of use, mission criticality, and embodied resources. Facilities program managers are essential participants in the analysis phase of life-cycle expenditures because they can ensure that operations and maintenance and repair expenditures over the projected lifetime of the facility are included as part of a facility's total cost.

The purpose of an extensive life-cycle cost analysis is to provide agency senior managers and Congress with the information they need to make informed decisions about the total costs of ownership before appropriating funds for new facilities. Accountability for the stewardship of facilities requires acknowledging the total costs of ownership and making a commitment to provide the funding necessary to maintain buildings over their entire life cycles.

Reducing the Number of Federal Facilities

Reducing the number of facilities owned and maintained by the federal government will result in substantial savings in operations and maintenance costs over the long term. It should also allow the government to better maintain facilities that are directly supportive of agency missions by redirecting available funding. Reducing the size of the portfolio will involve evaluating facilities to determine their relative importance or value to an agency, closing obsolete or underutilized facilities, transferring the ownership of viable but no longer needed facilities, and demolishing facilities that cannot be transferred or used effectively.

Agencies should conduct periodic analyses of their inventories to determine which properties no longer meet a standard of utility that warrants being retained. A facility may no longer be able to perform its original function for a variety of reasons. It may simply have reached the end of its useful life, it may have become technologically obsolete, or the function it was designed to house may have changed or become unnecessary. Facilities may also reach a point when it is more costly to operate and maintain them than to replace them.

One element of the process used to identify excess facilities could be similar in concept to a decision-making model used by private sector facility portfolio managers who periodically evaluate each property in an investment portfolio to determine whether it is dilutive (a negative asset) or accretive (a positive asset). The process involves developing an economic model of the portfolio; if the portfolio financial performance improves when the property is removed from the model, the property is dilutive and is a candidate for disposal. Accretive properties improve the portfolio and are candidates for retention. Other factors that play into the disposal/retention decision include long-range facility requirements (to ensure that existing facilities are not prematurely closed or transferred). Current programs and trends, as well as innovative approaches to meeting future mission requirements and trends in technology and industry and, in some cases, international competition, may also be considered.

Some factors are unique to the federal government. For instance, in a study of future needs for space facilities, the National Research Council found that

there are limits to the amount of consolidation of space facilities that can be undertaken without adverse effects on efficiency and capabilities. Some redundancy in R&D facilities is desirable to allow competition that spurs innovative thinking and to provide for contingencies. Some redundancy is also justified in operations facilities. For instance, the need for east- and west-coast launch capabilities will continue into the indefinite future, with obvious duplication of supporting infrastructure. . . . Attempts to modify old equipment to satisfy new requirements may respond to near-term budget constraints but be considerably more expensive in the long term and vastly less efficient (NRC, 1994).

Some properties can become obsolete for a particular agency although they may be perfectly serviceable for other agencies. When a facility is no longer valuable to an agency's portfolio or no longer contributes to the agency's mission, a simple and direct process of transferring title or otherwise disposing of the property would free maintenance and repair resources to be redirected to mission-critical facilities. The current process for declaring properties surplus and transferring title to other agencies or outside entities is cumbersome and time consuming. The government needs more efficient interagency communication about the potential availability of facilities. An agency that needs a new facility to implement its mission should be able to determine quickly and easily if an appropriate facility is available from another agency. Conversely, if an agency wants to dispose of a facility, it should be able to determine quickly if another agency could use the facility before making it available to nonfederal entities or taking it out of service. A centralized database for excess properties would make information sharing among government agencies more efficient and would expedite the disposition process.

If a facility is not needed by any federal agency, transferring ownership to a state or local public agency or the private sector should be considered. Transferring title of a facility will maximize the public's investment by reusing existing resources and will reduce federal operations and maintenance requirements.

The demolition of facilities that are functionally obsolete, do not support an agency's mission, are not historically significant, and are not suitable for transfer or adaptive reuse should also be part of the federal facilities portfolio management strategy. Up-front funding, over and above current operations and maintenance funding levels, should be considered for the demolition of facilities. The operations and maintenance funds that will be saved over the long term could then be redirected towards the maintenance and repair of mission-critical facilities.

Federal facilities support the provision of services, generate jobs, and are sometimes integral components in a community's architectural fabric. The transfer of title, closing, or demolition of a facility can generate considerable controversy at the local and congressional levels. Political and community pressures can make it difficult for agencies to transfer or close buildings even when they can clearly demonstrate that the facility is no longer needed. As the federal government continues to downsize and realign its services, more facilities, particularly civilian agencies' facilities, will be declared excess and either transferred to other entities or closed altogether. An independent, objective, outside panel may be needed to weigh the costs and benefits of transferring title or closing federal facilities and to build a political consensus for doing so. The Base Realignment and Closure Commission is a model that could be adapted for transferring or closing excess facilities owned by civilian agencies.

Strategy: Support Research on Maintenance and Repair-Related Issues

Facilities program managers could determine how maintenance and repair funds could be optimized and operate safer, healthy, and more productive facilities if they had access to information about cost-avoidance strategies, the deterioration of building components, and the nonquantitative effects of maintenance on agency mission and on the people who work in or rely on federal facilities. Research on cost-avoidance strategies and the deterioration of building components would help facilities program managers and others to plan and implement cost-effective facilities management programs and strategies, develop maintenance and repair budget requests, and determine the optimum time to repair or replace building components or systems to maximize service life and avoid business disruptions. Information on cost-avoidance strategies could also be used to convey the importance and cost effectiveness of preventive maintenance to the public and elected officials.

Research on the causes and "cures" of sick building syndrome would enable facilities managers to take measures to correct adverse building-related health factors that lead to absenteeism and lost productivity. A limited amount of research about how buildings can be designed and operated to enhance the working environment and productivity is under way. Additional studies are needed about the effects of timely maintenance, cost-avoidance strategies, cost analysis and cost estimating, and the deterioration of building components. Focused research in these areas would be useful for the maintenance not only of federal facilities but of all public and private buildings.

REFERENCES

COEQ and OMB (Council on Environmental Quality and Office of Management and Budget). 1995. Improving Federal Facilities Cleanup. Report of the Federal Facilities Policy Group. Washington, D.C.: Government Printing Office.

FFC (Federal Facilities Council). 1996. Budgeting for Facilities Maintenance and Repair. Technical Report No. 131. Standing Committee on Operations and Maintenance. Washington, D.C.: National Academy Press.

GAO (General Accounting Office). 1996. VA Health Care: Opportunities to Increase Efficiency and Reduce Resource Needs. Testimony. T-HEHS-96-99. Washington, D.C.: Government Printing Office.

GAO. 1997a. Budgeting Issues: Budgeting for Federal Capital. Chapter Report. Report to the Chair, Committee on Government Reform and Oversight, U.S. House of Representatives. AIMD-97-5. Washington, D.C.: Government Printing Office.

GAO. 1997b. Military Bases: Cost to Maintain Inactive Ammunition Plants and Closed Bases Could Be Reduced. Letter Report. NSIAD 97-56. Washington, D.C.: Government Printing Office.

GAO. 1997c. Defense Infrastructure: Demolition of Unneeded Buildings Can Help Avoid Operating Costs. Report to the Chair, Subcommittee on Military Installations and Facilities, Committee on National Security, U.S. House of Representatives. NSIAD-97-125. Washington, D.C.: Government Printing Office.

NPR (National Performance Review). 1993. From Red Tape to Results: Creating a Government That Works Better and Costs Less. Mission-Driven, Results-Oriented Budgeting. Washington, D.C.: Government Printing Office.

NRC (National Research Council). 1990. Committing to the Cost of Ownership: Maintenance and Repair of Public Buildings. Building Research Board, National Research Council. Washington, D.C.: National Academy Press.

NRC. 1993. The Fourth Dimension in Building: Strategies for Minimizing Obsolescence. Building Research Board, National Research Council. Washington, D.C.: National Academy Press.

NRC. 1994. Space Facilities: Meeting the Future Needs for Research, Development and Operations. Aeronautics and Space Engineering Board, National Research Council. Washington, D.C.: National Academy Press.

NRC. 1995. Measuring and Improving Infrastructure Performance. Board on Infrastructure and the Constructed Environment, National Research Council. Washington, D.C.: National Academy Press.

OMB (Office of Management and Budget). 1997. Analytical Perspectives, Budget of the United States Government, Fiscal Year 1998. Washington, D.C.: Government Printing Office.

5

Findings and Recommendations

The federal government owns and maintains more than 500,000 buildings and other constructed facilities to conduct the business of government and provide services to the public. More than 300 billion taxpayer dollars have been invested in acquiring these facilities, but relatively few resources are invested on an annual basis to ensure the functionality and quality of these facilities through effective management, maintenance, and repair. As a consequence, the GAO and other federal agencies report that the physical condition of the federal facilities portfolio is deteriorating and major repairs are required to bring many buildings up to acceptable safety, health, and performance levels. The GAO also reports that many necessary repairs were not made when they would have been most cost effective and are now part of a backlog of deferred maintenance.

The costs of ownership of a facility are equal to the total expenditures an owner makes over the course of a facility's service life, i.e., the costs of planning, design, construction, maintenance, repairs, normal operations, revitalization, and disposal. With proper management and maintenance, buildings may perform adequately for 40 to 100 years or more and may serve several different functions over that lifetime. Although a building's performance inevitably declines because of aging, wear and tear, and functional changes, its service life can be optimized through adequate and timely maintenance and repairs. Failure to provide adequate maintenance and failure to recognize the total costs of ownership results in a shorter service life, accelerated deterioration, and higher operating costs. The federal government's failure to recognize the total costs of ownership represents a lack of stewardship of the facilities themselves and of the public's investment in them.

Properly maintained federal facilities are not a luxury. They are critical to the effective performance of government agencies' missions and the provision of government services. Inadequate maintenance in public buildings can have serious and costly consequences. Damage caused by leaking roofs, burst pipes, and malfunctioning ventilation systems can disrupt work, cause computer and other technological breakdowns, create risks to occupants' health and safety, reduce productivity, and cost millions of dollars in emergency repairs.

The deferral of maintenance and repairs because of underfunding is a widespread, persistent, and long-standing problem, and pressures to defer maintenance are increasing. In today's dynamic policy and budget environment, federal facilities program managers are being challenged to extend the useful life of aging facilities; to alter or retrofit facilities to consolidate space or accommodate new functions and technologies; to meet evolving standards for safety, environmental quality, and accessibility; to maintain or dispose of underutilized, overutilized, and excess facilities; and to find innovative ways and technologies to maximize limited resources.

The specific findings of the committee regarding the state of the federal facilities portfolio and the practice of developing and implementing maintenance and repair budgets have been presented throughout the report and are summarized below.

FINDINGS

Finding 1. Based on the information available to the committee, the physical condition of the federal facilities portfolio continues to deteriorate, and many federal buildings require major repairs to bring them up to acceptable quality, health, and safety standards.

The deteriorating condition of federal facilities is due, in part, to the federal government's failure to recognize the total costs of facility ownership. Government budgeting practices are structured to focus on design and construction costs, which constitute only 5 to 10 percent of the total costs of ownership, rather than on the operations and maintenance of facilities, which account for 60 to 85 percent of total life-cycle costs. Thus, the emphasis has been on constructing and acquiring new buildings, rather than on maintaining, reusing, or leasing existing buildings.

Finding 2. The underfunding of facilities maintenance and repair programs is a persistent, long-standing problem. Federal operating and oversight agencies report that agencies have excess, aging facilities and insufficient funds to maintain, repair, or update them. Information provided to the committee indicated that agencies are receiving less than 2 percent of the aggregate current replacement value of their facilities inventories for maintenance and repair.

Because of inadequate funding agencies routinely defer maintenance, which can result in an irreversible loss of service life, the loss of functionality, and higher costs over time. Although there is no single, agreed upon guideline to determine the amount of money necessary to maintain public buildings effectively, an NRC report, Committing to the Cost of Ownership: Maintenance and Repair of Public Buildings, did recommend that, "An appropriate budget allocation for routine M&R [maintenance and repair] for a substantial inventory of facilities will typically be in the range of 2 to 4 percent of the aggregate current replacement value of those facilities" (NRC, 1990). This guideline has been widely quoted in the literature on facilities management. Variables that can have a major influence on the appropriate level of maintenance and repair expenditures include building size and complexity, age and condition, mechanical and electrical system technologies, telecommunications and security technologies, climate, and criticality of role or function, among others. Based on the information available to the committee, federal agencies receive less than 2 percent of the aggregate current replacement value of their facility inventories for routine maintenance and repair on an annual basis.

Finding 3. Federal government processes and practices are generally not structured to provide for effective accountability for the stewardship (i.e., responsible care) of federal facilities.

Because the decision-making responsibility related to federal facilities is delegated at all levels of government, no single entity can be held accountable for the results. Senior managers and public officials may think that they will not incur serious consequences if they defer the maintenance and repair of facilities for one more year in favor of more urgent operations or programs with greater visibility. Only if a roof falls in or there is a similar catastrophic failure, are agency managers likely to be held accountable for the condition of facilities in any given year. They are, however, held accountable for current operating programs. Consequently, they have few incentives to practice stewardship of the federal facilities portfolio, and they suffer few penalties if they do not.

Finding 4. Buildings and facilities are durable assets that contribute to the effective provision of government services and the fulfillment of agency missions. However, the relationship of facilities to agency missions has not been recognized adequately in federal strategic planning and budgeting processes.

Federal facilities embody significant assets and resources. Federal buildings and structures are acquired to support agencies in achieving their missions, to provide services to the public, and to provide workplaces for the people who conduct the government's business. The condition of federal facilities can affect an agency's ability to fulfill its mission, as well as the health and safety of occupants and building users. Evidence suggests that the physical condition and level of maintenance of buildings can also affect employees' productivity and morale and an agency's ability to recruit new staff. The only time agency officials and Congress discuss how facilities foster the implementation of an agency's mission is when reviewing budget requests for constructing or acquiring new facilities. Once a facility has been built, the relationship is taken for granted.

Finding 5. Maintenance and repair expenditures generally have less visible or less measurable benefits than other operating programs. Facilities program managers have found it difficult to make compelling arguments to justify these expenditures to public officials, senior managers, and budgeting staff.

In the federal budget and operations environment, facilities maintenance and repair is often considered a low priority issue because facilities program managers do not have the information they need to present an effective case for funding. In attempting to justify maintenance and repair budget requests, some federal agencies have kept inventories of building deficiencies and calculated the amount of funding it would take to eliminate the backlog of deficiencies, but public officials have not often found these justifications compelling.

Studies indicate that public officials do find arguments for the avoidance of future costs by early, preventive, or corrective maintenance more convincing and compelling. However, research to develop cost avoidance information and estimates of the costs of deferred maintenance in terms of money and quality of service have only been begun recently, and the results are not generally available.

Finding 6. Budgetary pressures on federal agency managers encourage them to divert potential maintenance and repair funds to support current operations, to meet new legislative requirements, or to pay for operating new facilities coming on line.

Federal agencies have some flexibility in allocating funding from their operations and maintenance accounts to either current operations or the maintenance and repair of facilities. There is considerable pressure on agency managers to allocate funding to current operations, for which they can be held accountable, instead of facilities, where accountability is difficult to assign. Other pressures on already limited maintenance and repair budgets arise through new legislative requirements to improve health, safety, or welfare that have facilities-related impacts. These requirements are usually enacted without the funding necessary to implement them (so-called unfunded mandates). Thus, removing hazardous materials or improving the accessibility of facilities must be funded from already limited agency operations and maintenance accounts. Although exact numbers on the costs of complying with these requirements are not available, anecdotal evidence clearly indicates that they have had an impact on operations and maintenance budgets and have resulted in the deferral of other maintenance and repair projects.

As new facilities come on line, funds to pay for their operation and maintenance must be allocated out of current operations and maintenance accounts. Thus, new facilities also create pressure on managers to divert funds that might otherwise be used to maintain existing buildings.

Finding 7. It is difficult, if not impossible, to determine how much money the federal government as a whole appropriates and spends for the maintenance and repair of federal facilities because definitions and calculations of facilities-related budget items, methodologies for developing budgets, and accounting and reporting systems for tracking maintenance and repair expenditures vary.

The methodologies used to formulate maintenance and repair budget requests vary from one agency to another. Accounting structures and the definitions of elements in those accounting structures also vary from agency to agency. As a result, direct comparisons of maintenance and repair allocations and expenditures across federal agencies are difficult to make.

Because maintenance and repair funds in most agencies are included in the operations account, they are not "earmarked" for specific maintenance and repair activities. Structuring the account this way blurs the line between maintenance and repair work, operations, and alterations and provides federal agencies with considerable flexibility in determining how much funding to allocate to maintenance and repair activities and which projects to fund. A detailed cost accounting showing the amount of funding actually appropriated to maintenance and repair activities is not required, and therefore few, if any, agencies complete one.

Finding 8. There is evidence that some agencies own and are responsible for more facilities than they need to support their missions or than they can maintain with current or projected budgets.

The federal facilities portfolio has grown over time in response to new programs and requirements, defense and foreign policy initiatives, changing demographics, and other factors. Little emphasis has been placed on demolishing obsolete facilities or divesting the government of no longer needed, but still viable, properties. Consequently, some agencies own buildings and properties that are no longer used or are otherwise underutilized but which they are still responsible for maintaining. The creation of excess federal facilities has accelerated as agencies have realigned their missions in response to changing circumstances. The most dramatic, but not the only, example is the Base Realignment and Closure process, through which one of every five military installations is slated to be closed.

Finding 9. Federal facilities program managers are being encouraged to be more businesslike and innovative, but current management, budgeting, and financial processes have disincentives and institutional barriers to cost-effective facilities management and maintenance practices.

The federal budget is a unified, cash-based budget that treats outlays for capital and operating activities the same way. This process is inherently biased against capital projects because the budget makes no distinction between outlays for capital assets that produce future benefits and outlays for current operations. Because capital projects tend to require relatively large outlays of money in the short run, they are often foregone to meet short-term budget restraints despite their long-term benefits.

The budget process also discourages cost-effective maintenance by disallowing, in most circumstances, the carryover of unobligated funds from one fiscal year to the next even if a facilities program manager can demonstrate that carryover funding is the most cost-effective way to fund a capital improvement. Funds that are not expended in the current fiscal year are routinely taken back from the agencies, and the next fiscal year's funding may be reduced on the premise that money not spent is money not needed. Thus, "admitting to savings" is not in a manager's interest. Because of the absence of rewards for cost-effective, fiscally responsible management, facilities program managers have few incentives to act in innovative ways or to take risks that might lead to more cost-effective maintenance and repair programs and strategies.

Finding 10. Performance measures to determine the effectiveness of maintenance and repair expenditures have not been developed within the federal government. Thus, it is difficult to identify best practices for facilities maintenance and repair programs across or within federal agencies.

Simply knowing how much money and staff time were allocated to maintenance and repair programs does not indicate how effectively those resources were used. Because government agencies do not consistently track maintenance and repair expenditures, it is difficult to develop measures to determine how effectively funds are being spent either within or across agencies. Thus, facilities program managers find it difficult to determine how effectively maintenance and repair funds are used and have been unable to develop benchmarks by which to identify "best practices" for facilities management and maintenance and repair across federal agencies.

Finding 11. Based on the information available to the committee, federal condition assessment programs are labor intensive, time consuming, and expensive. Agencies have had limited success in making effective use of the data they gather for budget development or for the ongoing management of facilities.

Information gathered from condition assessments can be used to (1) estimate maintenance and repair needs; (2) develop cost estimates and funding priorities for various projects; and (3) generate and prioritize work orders. Condition assessment programs in federal agencies are evolving from simple catalogues of maintenance and repair deficiencies to computerized programs with automated checklists that link condition assessment data to agency mission and improve facility management. Federal condition assessment programs have generally been developed independently to meet the needs of individual agencies within their funding constraints. As a consequence, the level of sophistication varies widely. Agencies with condition assessment programs gather a wide range of data, including cosmetic problems, structural problems, and mechanical deficiencies. Because of the amount of data being collected and the time and resources required to analyze it, federal agencies have had limited success in using condition-related data to support ongoing facilities management or to develop maintenance and repair budget requests.

Finding 12. Organizational downsizing has forced facilities program managers to look increasingly to technology solutions to provide facilities-related data for decision making and for performing condition assessments.

Technology related to facilities management and inspections is evolving rapidly. In recent years, progress has been made in the development of CAFMs (computer aided facility management systems) and CMMS (computerized maintenance management systems) although few standards have been established. Technologies, such as pen-based data collection devices, bar code scanners, and digital and video cameras for condition assessments which have been deployed also have some drawbacks. Nondestructive evaluation technologies for detecting building deficiencies are being used by a few agencies in limited situations.

Finding 13. Existing sensor and microprocessor technologies have the potential to monitor and manage a range of building conditions and environmental parameters, but, for economic and other reasons, they have not been widely deployed.

Today's intelligent buildings integrate sensor and monitoring devices, data transmission via telephone lines, fiber optic cable or satellite uplinks, computers for data management and decision making, and microprocessor control devices for various types of mechanical equipment. Although currently available intelligent building technologies have many possible applications for the monitoring and assessment of buildings and building systems, proactive diagnostic systems have not been widely deployed. The primary reason appears to be the lack of well documented economic paybacks to justify the initial costs. Also hampering their deployment is the lack of a standard protocol for communications among the various devices.

Finding 14. Training for staff is a key component of effective decision making, condition assessments, and the development of maintenance and repair budgets.

Because of reductions in total staffing levels, record numbers of experienced personnel have left the government. The loss of their technical expertise has significant implications for the maintenance and repair of federal facilities. Automating the condition assessment process will require different skills, particularly computer skills, than are typically found in facilities management organizations. A lack of sensitivity within the budget process to the total costs of facilities ownership is a key factor in the long-term underfunding of maintenance and repair programs and lack of technical expertise will exacerbate this problem.

Finding 15. Only a limited amount of research has been done on the deterioration/failure rates of building components or the nonquantitative implications of building maintenance (or lack thereof). This research is necessary to identify effective facilities management strategies for achieving cost savings, identifying cost avoidances, and providing safe, healthy, productive work environments.

Research on facility management related issues has only begun recently. No standard methods have been developed for estimating the future costs of deferred maintenance for a particular facility. Relatively little research has been done on the deterioration rates of building components, which are essential to estimating cost-avoidance strategies. Predictions of deterioration/failure rates would also be useful for estimating future budget needs, determining the optimal repair/replacement cycles for particular types of infrastructure, and for analyzing life-cycle building costs. Current assessments of the nonquantitative effects of poorly maintained buildings, such as reduced mission delivery or the effects on employees' health, safety, and welfare, are based primarily on qualitative, subjective judgments rather than on empirical data.

Finding 16. Greater accountability for the stewardship of facilities is necessary at all levels of the federal government. Accountability includes being held responsible for the condition of facilities and for the allocation, tracking, and effective use of maintenance and repair funds.

Those responsible for making decisions on the funding, acquisition, maintenance and repair, and disposition of federal facilities and other facilities-related activities, include Congress, oversight agencies, senior executives, program managers, field engineers, and others. Because the responsibility for facilities-related decisions is spread throughout the government, no single entity can be held accountable for the results.

Based on these findings, the committee developed the following recommendations aimed at fostering facility stewardship, including effective strategic planning, improved budgetary techniques and processes, wider deployment of technology, and the development of necessary staff skills. The committee recommends that the following actions, which are not in any particular order of priority, be taken.

RECOMMENDATIONS

Recommendation 1. The federal government should plan strategically for the maintenance and repair of its facilities in order to optimize available resources, maintain the functionality and quality of federal facilities, and to protect the public's investment. A recommended strategic framework of methods, practices, and strategies for the proactive management and maintenance of the nation's public assets is summarized in Figure 5-1 (Findings 1 and 2).

Maintenance and repair of federal facilities is a complex issue, and no single action or strategy will resolve all of the associated problems. All levels of the federal government will have to make a commitment to solve these problems over the long term to optimize resources and sustain the public's investment in the facilities portfolio.

Recommendation 2. The government should foster accountability for the stewardship of federal facilities at all levels. Facilities program managers at the agency level should identify and justify the resources necessary to maintain facilities effectively and should be held accountable for the use of these resources (Findings 1, 2, 3 and 16).

Buildings, other constructed facilities, and associated infrastructures represent hundreds of billions of dollars in assets and resources and support the effective provision of government services. Adequately maintained facilities are critical to the achievement of agency missions and to organizational and individual performance. Senior agency managers should strive to create a climate of stewardship as their basic business strategy.

Recommendation 3. At the executive level, an advisory group of senior level federal managers, other public sector managers, and representatives of the nonprofit and private sectors should be established to develop policies and strategies to foster accountability for the stewardship of facilities and to allocate resources strategically for their maintenance and repair (Findings 1, 2, 3 and 16).

Stewardship of the federal facilities portfolio involves exercising responsible care over the facilities investment, including maximizing the use of facilities, optimizing service life and building performance, and sustaining the quality and functionality of facilities through reinvestment. Fostering accountability for the stewardship of federal facilities at the highest levels of government is at least as important as fostering accountability at the agency and field-office level. Senior leadership should provide guidance for responsible, cost effective ways to manage the federal facilities portfolio and to protect the public's investment. An executive level federal facilities advisory group should be appointed to increase the visibility of the issue of federal facilities maintenance, repair, and stewardship.

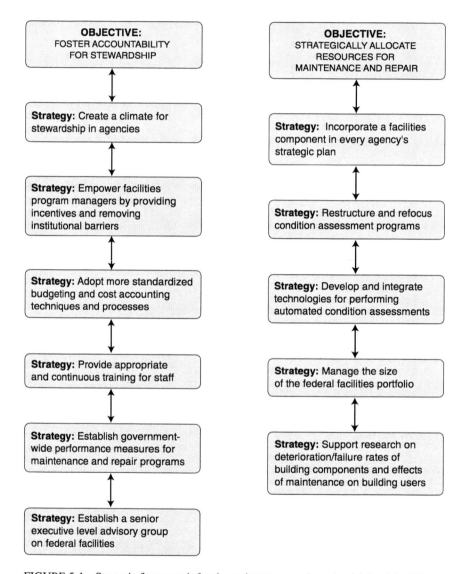

GOAL
Protect and Enhance the Functionality and Quality
of the Federal Facilities Portfolio

OBJECTIVE:
FOSTER ACCOUNTABILITY
FOR STEWARDSHIP

OBJECTIVE:
STRATEGICALLY ALLOCATE
RESOURCES FOR
MAINTENANCE AND REPAIR

Strategy: Create a climate for stewardship in agencies

Strategy: Incorporate a facilities component in every agency's strategic plan

Strategy: Empower facilities program managers by providing incentives and removing institutional barriers

Strategy: Restructure and refocus condition assessment programs

Strategy: Adopt more standardized budgeting and cost accounting techniques and processes

Strategy: Develop and integrate technologies for performing automated condition assessments

Strategy: Provide appropriate and continuous training for staff

Strategy: Manage the size of the federal facilities portfolio

Strategy: Establish government-wide performance measures for maintenance and repair programs

Strategy: Support research on deterioration/failure rates of building components and effects of maintenance on building users

Strategy: Establish a senior executive level advisory group on federal facilities

FIGURE 5-1   Strategic framework for the maintenance and repair of federal facilities.

The advisory group should provide policy direction and set priorities for the management and maintenance of the facilities portfolio.

Recommendation 4. Facility investment and management should be directly linked to agency mission. Every agency's strategic plan should include a facilities component that links facilities to agency mission and establishes a basis and rationale for maintenance and repair budget requests (Finding 4).

The Government Performance and Results Act of 1993 requires that federal agencies develop strategic plans. As agencies reevaluate and, in some cases, redefine their missions, the relationship between agency mission and the facilities that support the mission should be made explicit. Beyond simply including facilities in the strategic planning process, strategic plans should contain a facilities component that recognizes facilities that are critical to achieving the agency's mission and point up the need for allocating resources to maintain them at an appropriate level of performance. Linking facilities to mission explicitly will enable agencies to link maintenance and repair budget requests and allocations to their long-term strategic planning and to the strategic plans of the government as a whole.

Recommendation 5. The government should adopt more standardized budgeting and cost accounting techniques and processes to facilitate tracking of maintenance and repair funding requests, allocations, and expenditures and reflect the total costs of facilities ownership. The committee developed an illustrative template as shown in Figure 5-2 (Findings 3, 5, 6, 7 and 16).

Greater standardization is needed in budgeting, cost accounting, definitions of activities, and calculations of facilities-related terms to foster accountability for the allocation, tracking, and effective use of maintenance and repair funds. The illustrative template developed by the committee is intended to clearly identify the total costs of facilities ownership. This template could serve as the basis for a more detailed standardized chart of accounts. If most or all government agencies adopted this template for formulating budget requests and tracking allocations and expenditures, in conjunction with their existing systems, budget expenditures could be compared and standardized measures developed. These measures could then be used to develop benchmarks and identify best practices for facilities management and maintenance.

Recommendation 6. Government-wide performance measures should be established to evaluate the effectiveness of facilities maintenance and repair programs and expenditures (Finding 10).

Unlike budgeting procedures and practices, performance measurements are not yet well developed or ingrained in federal agency practices and procedures.

Facilities Management-Related Activities	Included in 2-4% Benchmark	Funding Category and Comments
A. Routine Maintenance, Repairs, and Replacements • recurring, annual maintenance and repairs including maintenance of structures and utility systems, (including repairs under a given $ limit, e.g., $150,000 to $500,000 exclusive of furniture and office equipment) • roofing, chiller/boiler replacement, electrical/lighting, etc. • preventive maintenance • preservation/cyclical maintenance • deferred maintenance backlog • service calls	Yes	Annual operating budget
B. Facilities-Related Operations • custodial work (i.e., services and cleaning) • utilities (electric, gas, etc./plant operations) • snow removal • waste collection and removal • pest control • security services • grounds care • parking • fire protection services	No	Annual operating budget
C. Alterations and Capital Improvements • major alterations to subsystems, (e.g., enclosure, interior, mechanical, electrical expansion) that change the capacity or extend the service life of a facility • minor alterations (individual project limit to be determined by agency $50,000 to $1 million)	No	Various funding sources, including no year, project-based allocations such as revolving funds, carryover of unobligated funds, funding resulting from cost savings or cost avoidance strategies
D. Legislatively Mandated Activities • improvements for accessibility, hazardous materials removal, etc.	No	Various sources of funding
E. New Construction and Total Renovation Activities	No	Project-based allocations separate from operations and maintenance budget. Should include a life-cycle cost analysis prior to funding
F. Demolition Activities	No	Various sources of funding

FIGURE 5-2   Illustrative template to reflect the total costs of facilities ownership.

Agencies have an opportunity to develop performance measures that can be used consistently across the government to compare the effectiveness of facilities management and maintenance programs. Uniform measures are particularly important in organizational structures with decentralized functions, in which individual centers and installations have discretion over how funds are spent.

Recommendation 7. Facilities program managers should be empowered to operate in a more businesslike manner by removing institutional barriers and providing incentives for improving cost-effective use of maintenance and repair funds. The carryover of unobligated funds and the establishment of revolving funds for nonrecurring maintenance needs should be allowed if they are justified (Findings 3 and 9).

Employees need incentives to work toward improvement or to take risks that could (or could not) result in cost savings. If facilities program managers are to be held accountable for the consequences of their actions, they should be given the appropriate tools, funding, and authority to carry out their responsibilities. Potential rewards for facilities program managers include allowing them to take savings from one area of operations and maintenance and apply it to another; allowing them to carry over unobligated funds from one fiscal year to the next for capital improvements, if this can be shown to be cost effective; or establishing awards for facilities maintenance and repair programs with high levels of performance. Revolving funds offer potential advantages for addressing maintenance and repair needs. To provide for accountability, the actual savings achieved through the implementation of any of these or other strategies should be well documented and should appear in the budget, which should also specify how the savings will be used.

Recommendation 8. Long-term requirements for maintenance and repair expenditures should be managed by reducing the size of the federal facilities portfolio. New construction should be limited, existing buildings should be adapted to new uses, and the ownership of unneeded buildings should be transferred to other public and private organizations. Facilities that are functionally obsolete, are not needed to support an agency's mission, are not historically significant, and are not suitable for transfer or adaptive reuse should be demolished when it is cost effective to do so (Findings 2, 8 and 16).

As a result of decisions made over many decades, some agencies in the federal government now own more facilities than they need to conduct their business. Responsible stewardship requires that the size of the facilities portfolio be managed effectively and reduced to a level that supports the long-term mission of the government and its agencies (1) by limiting the construction and acquisition of new facilities; and (2) by reducing the total number of facilities owned and maintained by the federal government. Reducing the total number of facilities will result in substantial savings in operations and maintenance costs over the long term and allow agencies to redirect funds to facilities that directly support their missions.

The transfer of title, closing, or demolition of facilities can generate considerable controversy at the local level and in Congress. An independent, objective, outside panel may be necessary to weigh the costs and benefits of transferring or closing federal facilities and to build a political consensus for doing so.

Recommendation 9. Condition assessment programs should be restructured to focus first on facilities that are critical to an agency's mission; on life, health, and safety issues; and on building systems that are critical to a facility's performance. This will optimize available resources, provide timely and accurate data for formulating maintenance and repair budgets, and provide critical information for the ongoing management of facilities (Findings 4 and 11).

To optimize the value of condition assessments, agencies should consider restructuring them. Data gathering should focus on information that is critical to building performance, building users' safety and health, and informed decision making. Condition assessments should concentrate on critical elements that affect the ability of an agency to operate effectively rather than simply cataloguing problems that may give relatively equal weight to structural and cosmetic deficiencies. Linking condition assessments to mission-critical facilities first and focusing on life, safety, and health standards and critical building system components should help integrate them into the strategic planning and budgeting processes of federal agencies.

Recommendation 10. The government should provide appropriate and continuous training for staff that perform condition assessments and develop and review maintenance and repair budgets to foster informed decision making on issues related to the stewardship of federal facilities and the total costs of facilities ownership (Findings 14 and 16).

A firm grounding in the principles of facilities management and an understanding of the relationship between adequate and timely maintenance and repair to total costs of facilities ownership are critical for anyone charged with the preparation or review of facilities management budgets. Staff having these responsibilities should be trained in the principles of facilities management and related topics and this training should be updated on a continuous basis.

The increasing number of intelligent buildings with building automation systems requires facility personnel who are familiar with a broad range of computer applications (e.g., graphics, databases, and spreadsheets) as well as hardware (e.g., personal computers and microprocessors). Personnel involved in automated condition assessments should have a similar level of computer skills. As buildings become more sophisticated, staff will have to maintain and update software systems which will require essentially continuous training for operations and maintenance personnel.

Recommendation 11. The government and private industry should work together to develop and integrate technologies for performing automated facility condition assessments and to eliminate barriers to their deployment (Findings 11, 12 and 13).

Automating the condition assessment process has the potential for cost savings, improved building performance, and a means of coping with reduced staffing levels. The data necessary to test these assumptions could be obtained either by designing and installing automated condition assessment systems in new federal buildings or by studying and evaluating buildings that already have them in the private sector. The federal government's responsibility for the long-term stewardship of buildings and facilities supports this kind of leadership position in the deployment of new building technologies and the acceptance of higher first costs to reduce life-cycle costs.

Recommendation 12. The government should support research on the deterioration/failure rates of building components and the nonquantitative effects of building maintenance (or lack thereof) in order to develop quantitative data that can be used for planning and implementing cost-effective maintenance and repair programs and strategies and for better understanding the programmatic effects of maintenance on mission delivery and building users' health, safety, and productivity (Findings 12 and 15).

To improve the management of facilities, to determine how maintenance and repair funds can be optimized, and to present budget requests effectively to senior agency managers and public officials, facilities program managers need access to more information about maintenance and repair cost-avoidance strategies and the deterioration of building components. This information would help them determine when individual components or systems should be repaired or replaced and how maintenance should be timed to optimize service life and minimize business disruptions. Information about cost avoidance is critical for conveying the importance and cost effectiveness of preventive maintenance to elected officials and the public.

REFERENCE

NRC (National Research Council). 1990. Committing to the Cost of Ownership: Maintenance and Repair of Public Buildings. Building Research Board, National Research Council. Washington, D.C.: National Academy Press.

APPENDIX

Biographical Sketches of Committee Members

Jack E. Buffington, chair, was elected to the National Academy of Engineering in 1996. RADM Buffington (CEC U.S. Navy, retired) currently heads the Mack-Blackwell National Rural Transportation Study Center at the University of Arkansas Department of Civil Engineering, where he is responsible for directing studies by professors and students to improve life in rural America through improvements in transportation systems. Admiral Buffington served for 34 years with the Naval Facilities Engineering Command (NAVFAC), rising to commander and chief of civil engineers. He led a team of 22,000 NAVFAC employees, with an annual workload of $7 billion, to provide engineering and contracting support for environmental, design, construction, and public works operations worldwide. His previous positions with NAVFAC included commander of the Pacific Division, commanding officer of the Navy Public Works Center in Norfolk, Virginia, officer in charge of construction of the $200 million Bethesda Naval Medical Center complex, and commander of both the Atlantic and Pacific Seabees. Admiral Buffington is the past national president of the Society of American Military Engineers and was elected to the Board of Directors of the National Institute of Building Sciences in 1996. He holds B.S. and M.S. degrees in civil engineering from the University of Arkansas and the Georgia Institute of Technology, respectively.

Albert F. Appleton is a senior fellow with the Regional Plan Association (RPA), the oldest regional planning organization in the United States. Mr. Appleton has developed regional infrastructure and environmental financing strategies and programs for RPA's new Third Regional Plan, which is built around integrating engineering and environmental expertise and linking investments in infrastructure, environmental, economic, financial, and community renewal to increase public benefits and reduce costs. He also is a consultant on infrastructure policy and privatization, both nationally and internationally. Mr. Appleton served on the National Research Council Synthesis Committee for the Study of Sustainable Habitats. Before joining the RPA, Mr. Appleton was the commissioner of the New York City Department of Environmental Protection, where he carried out a major restructuring of the department's management, finances, and programs. Under his tenure, the department set records for capital construction and established modern business and financial planning systems. He was also executive assistant attorney general for the New York State Medicaid Fraud Control Unit and senior project planner in the Criminal Justice Coordinating Council of the Mayor of New York City. Mr. Appleton holds a B.A. from Gonzaga University and an LL.B. from Yale University Law School.

Gary G. Briggs is senior vice president and chief operating officer of Consolidated Engineering Services, Inc. He founded this diversified engineering company from components of the Charles E. Smith Companies to provide technical and consulting services for operations, maintenance, and facilities management, including mechanical plants and operations, fire safety and security, elevators and escalators, structures and envelopes, and energy management. These services are provided to more than 100 properties, including office buildings, residential units, retail malls, and recreational facilities. Previously, Mr. Briggs was the senior vice president and head of the Mechanical Department of Charles E. Smith Management, Inc. He has served on several National Research Council committees, including the Committee on Feasibility of Applying Blast-Mitigating Technologies and Design Methodologies from Military Facilities to Civilian Facilities, the Committee on Facility Design to Minimize Premature Obsolescence, and the Committee on Heating, Ventilating, and Air Conditioning Criteria. He holds a B.S. in physics from Drexel University and is a member of the American Society of Heating, Refrigerating and Air Conditioning Engineers, the Building Owners and Managers Association, and the International Facility Management Association, among other professional organizations.

Sebastian J. Calanni, retired, was the senior vice president for federal government services for Johnson Controls, Inc., a leading provider of facilities management, operations, maintenance, technical, and institutional services to government and commercial organizations. Mr. Calanni had general management responsibilities for all federal-sector business of Johnson Controls' integrated facility management group. In this position, he provided corporate oversight for more than 40 federal government facility and infrastructure contracts valued at more than $500 million per year and involving more than 7,000 employees. In previous positions with Pan Am World Services, Mr. Calanni was project director for plant maintenance and operations at NASA's Johnson Space Center, and area manager for the Electromagnetic Environmental Test Facility at Fort Huachuca, Arizona, for the U.S. Army Signal Corps. Mr. Calanni has a B.S. in industrial management and an M.B.A. from the University of Houston. He is the recipient of three NASA Group Achievement Awards and participates in many civic and community organizations.

Eric T. Dillinger is the director of facilities management services with Carter & Burgess, Inc., a consulting firm that provides multidisciplinary engineering, architectural, planning, and surveying services. Mr. Dillinger's primary focus over the past 10 years has been in the area of facility management, including facility audits/condition assessment surveys, resource allocation, and capital asset management. He has participated in and directed facility audits and capital asset management programs for numerous federal government installations and agencies, as well as private-sector organizations. Mr. Dillinger also has extensive experience in architectural/engineering endeavors, maintenance and repair prioritization, preventive and predictive maintenance, space utilization, inventory control, and scheduling and resource programming. His experience includes capital asset management for more than 12.5 million square feet of facilities and an operations and maintenance budget of more than $25 million per year. Mr. Dillinger was the primary author of the Component Identification and Inspection Evaluation Standards Manual (1990) and Value Based Budgeting (1991) for the U.S. government. He also contributed to A Practical Guide to Neural Nets (1991). He has conducted seminars and trained facility managers, inspectors, and resource managers in facility maintenance standards, resource allocation, prioritization, and capital asset management for both government and private-sector clients. He has a B.S. in industrial engineering from Kansas State University and is a member of the International Facility Management Association, the Association of Higher Education Facility Officers, the Society of American Military Engineers (past post president), and the Society of American Military Comptrollers, among other organizations.

William L. Gregory is manager of Corporate Facilities Management at Kennametal, Inc., a global provider of industrial tooling systems with annual sales of nearly $1 billion and 7,000 employees worldwide. At Kennametal, Mr. Gregory is responsible for real estate, corporate building operations, strategic facility planning, corporate environmental and health and safety programs, space management, installations, and construction management for all major facility projects on a global basis. Recent major projects include the construction of a manufacturing plant in China, a corporate technology center for research and development, and a new corporate headquarters. Mr. Gregory has also worked for Kennametal as a project manager, a facility manager, and a project electrical engineer. In a previous position with Westinghouse Electric Corporation, he was a project engineer. Mr. Gregory is a past international president of the International

Facilities Management Association (1993-1994) and past international vice president, treasurer, and regional vice president for the northeast region. As president of the association, Mr. Gregory oversaw its operation, including its international development and the formation of professional alliances. He has a B.S. in electrical engineering from Grove City College. He is a certified facility manager and a registered engineer in Pennsylvania and Ohio.

B. James Halpern is president and chairman of the Board of Directors of Measuring and Monitoring Services, Inc. (MMSI), which specializes in field monitoring, data acquisition, information processing, and reporting for a variety of applications in the energy, water, and environmental industries. MMSI also designs, develops, and manufactures integrated systems for comprehensive data acquisition and reporting. Mr. Halpern created a product line and founded MMSI to provide measurement services to the energy services industry, focusing on demand-side management and utilizing performance-based contracts. Mr. Halpern is currently chairman of the U.S. Department of Energy's Technical Subcommittee for the Establishment of a National Energy Measurement and Verification Protocol. Previously he was president of Energy Futures, a company he created to market and develop alternatively financed energy projects in New Jersey. He also was president of REEP, Inc., a residential conservation company that serviced utilities throughout the mid-Atlantic region. Mr. Halpern has a B.A. in architecture from Carnegie Mellon University. He is a member of the Association of Energy Engineers and the Illuminating Engineer Society of North America, among other organizations.

James E. Kee is the interim dean and professor of public administration in the School of Business and Public Management at the George Washington University. He is also the Giant Food, Inc., Professor of Public/Private Management. As dean, his responsibilities include budgeting, planning, and faculty development for a school with 120 full-time faculty members, 3,000 students, a budget of $15 million, and revenues of more than $31 million. He is also responsible for the development and funding of research centers and the development and implementation of the school's strategic plan. He is the lead professor in the fields of budgeting and public finance and managing state and local governments. Professor Kee has also been the senior associate dean, the chair of the Department of Public Administration, a faculty associate in Public Policy, and a member of the University President's Budget Advisory Team. His teaching and research interests include public expenditure analysis, budget and tax policy, intergovernmental relations and finance, state and local government management and finance, and developing and maintaining organizational excellence. Professor Kee has written chapters in several books dealing with benefit-cost analysis and strategic management in the federal government and has written numerous articles on public finance for all levels of government. He was the executive director of the

Department of Administrative Services for the State of Utah from 1981 to 1985, where he was responsible for the coordinated management of finances and administrative services. During his tenure, a new on-line financial information management system was created, and the annual $100 million capital building program and procedures were streamlined. He also served as the Utah state budget director, where he developed the state's first capital budget in 1981, and as state planning coordinator from 1976 to 1978. He has a B.A. from the University of Notre Dame, a J.D. from New York University School of Law, and an M.P.A. from New York University.

Vivian E. Loftness is a professor of architecture at the Center for Building Performance and Diagnostics, head of the Department of Architecture, College of Fine Arts at Carnegie Mellon University, and a registered architect. She is an international consultant on energy and building performance for commercial and residential building design and has researched and written extensively on building performance, energy conservation, and design-related subjects. Professor Loftness is conducting advanced architectural research in the performance of a range of building types, from museums to high-tech offices, and on innovative building delivery processes for improving the quality of building performance. Supported by the Advanced Building Systems Integration Consortium, a university-building industry partnership, Ms. Loftness has been actively researching and designing high performance office environments and was a key contributor to the creation of the Intelligent Workplace, a living laboratory of commercial building performance innovations. Ms. Loftness has authored a range of publications on international advances in the workplace. She has B.S. and M.S. degrees in architecture from the Massachusetts Institute of Technology. Professor Loftness is a member of the National Research Council (NRC) Board on Infrastructure and the Constructed Environment and has previously served on several NRC committees, including the Committee on Advanced Maintenance Concepts in Buildings and the Committee on Electronically Enhanced Buildings. She was also a consultant to the Committee on Building Diagnostics.

Terrance C. Ryan is the assistant dean and professor of urban systems engineering in the Civil, Environmental, and Infrastructure Engineering Program at George Mason University (GMU). Dr. Ryan received M.S. and Ph.D. degrees in civil engineering systems from the University of Illinois, the latter after several years of field engineering experience with the U.S. Army Corps of Engineers. He is a graduate of the U.S. Military Academy and has more than 30 years of experience in construction management, information technology, operations research, and teaching. Dr. Ryan joined the faculty of GMU in 1989 from George Washington University, where he was a distinguished visiting professor and taught construction management and decision science. While working at the U.S. Army's Construction Engineering Research Laboratories, Dr. Ryan was named Researcher of the Year, and he has maintained his research interests in the applications of information technology and decision science in engineering and construction. He has been a construction manager of a large academic facility built for the Saudi Arabian government. As a vice president for a specialty retail chain, he was responsible for the design, construction, and maintenance of more than 120 stores and 300,000 square feet of space. A registered professional engineer in Virginia, his consulting experience has been in small business management, project management, and as regional director for a minority-owned architect/engineer firm specializing in project management services. Dr. Ryan is currently the co-director of the Facilities Management Laboratory, executive director of the Urban Systems Engineering Institute, and director of the Fairfax/GMU Center for Community Reinvestment at George Mason University.

Richard L. Siegle is director of facilities for the Washington State Historical Society, where he is responsible for the planning, design, construction, and management of facilities owned by the society. His recent responsibilities include oversight of the development of the new $42 million state history museum. From 1986 to 1995, Mr. Siegle was responsible for the planning, design, construction, and maintenance of all Smithsonian Institution museum and research facilities. Resources included 1,290 professional and support staff and an annual operating budget of $125 million. Capital projects during the period exceeded $500 million. His responsibilities included preparing capital and facilities operating budgets for House and Senate committees of the Congress. Mr. Siegle was also a contributing member of the Board and Design Committee of the Pennsylvania Avenue Development Corporation, which was involved in creating facilities in excess of $750 million. He was the director of design and construction of state buildings and the deputy director for the Department of General Administration for the State of Washington from 1978 to 1986. As an officer for more than 20 years with the Navy Civil Engineer Corps, Mr. Siegle served in engineering and teaching positions throughout the United States and in the Pacific and Far East. He is a registered professional engineer, a fellow of the American Society of Civil Engineers, and a member of the National Society of Professional Engineers, the Association of Physical Plant Administrators, and the American Association of Museums. He previously served on the National Research Council Committee on Infrastructure. Mr. Siegle received a B.S. in civil engineering from the University of Illinois and an M.S. in civil engineering from Stanford University.

George M. White was appointed vice chairman of Leo A. Daly, one of the oldest and largest multidisciplinary design and management firms in the United States, in 1996. He previously served as the Architect of the Capitol for 25 years, where he was responsible for overseeing 13 million square feet of federal space, 2,300 employees, and an annual budget of approximately $180 to $200 million. Mr. White oversaw the restoration of the old Supreme Court chamber, the old Senate chamber and the west front of the U.S. Capitol, the construction of the Library of Congress James Madison Memorial Building, the Hart Senate Office Building, the extension of the U.S. Capitol Power Plant, and the Thurgood Marshall Federal Judiciary Building, among other projects. He has also practiced architecture and law in the private sector and worked as a design engineer with the General Electric Company. Mr. White holds B.S. and M.S. degrees from the Massachusetts Institute of Technology, an M.B.A. from Harvard University, and a J.D. from Case Western Reserve University. He is a fellow of the American Institute of Architects, an honorary fellow of the American Society of Civil Engineers, and a member of the National Society of Professional Engineers. He has also served on numerous panels and commissions and is the recipient of many awards.

Bibliography

American Public Works Association. 1992. Plan. Predict. Prevent. How to Reinvest in Public Buildings. Special Report #62. Chicago: American Public Works Association.

Applied Management Engineering. 1991. Managing the Facilities Portfolio: A Practical Approach to Institutional Facility Renewal and Deferred Maintenance. Washington, D.C.: National Association of College and University Business Officers.

American Society of Heating, Refrigeration, and Air Conditioning Engineers. 1994. BACnet - A Data Communication Protocol for Building Automation and Control Networks. Document SPC-135P-031. Atlanta, Ga: American Society of Heating, Refrigeration, and Air Conditioning Engineers.

ASTM (American Society for Testing and Materials). 1992. Standards on Building Economics. Philadelphia, Pa.: American Society for Testing and Materials.

ASTM. 1997. Standards on Whole Building Functionality and Serviceability. Philadelphia, Pa.: American Society for Testing and Materials.

Ault, Douglas K., and William A. Woodring. 1985. Funding Requirements for Maintenance of Real Property. Bethesda, Md: Logistics Management Institute.

Barco, A.L. 1994. Budgeting for facility repair and maintenance, Journal of Management in Engineering 10(4): 28–34.

Bowyer, Robert A. 1993. Capital Improvements Programs: Linking Budgets to Planning. Washington, D.C.: American Planning Association Press.

Broom, Cheryle A. 1995. Performance-based government models: building a track record, Public Budgeting and Finance 15(4): 3–17.

Cable, John, Mark A. Fritzlen, Jeffrey S. Frost, and Thomas P. Wilson. 1996. Use of Information Technology for Management of U.S. Postal Service Facilities. McLean, Va.: Logistics Management Institute.

Cable, John, Marguerite Moss, and Adam Dooley. 1995. Repair and Alteration Services at the U.S. Postal Service. McLean, Va.: Logistics Management Institute.

Carter, John P. and Richard Mayo. 1994. Prediction of repair parts budget requirement, Cost Engineering 36(4): 15–22.

Chouinard, L.E., G.R. Andersen, and V.H. Torrey, III. 1996. Ranking models used in condition assessment of civil infrastructure systems, Journal of Infrastructure Systems 2(1): 23–29.

Christian, John, and Amar Pandeya. 1997. Cost predictions of facilities, Journal of Management in Engineering. New York: American Society of Civil Engineers 13(1): 52–61.

Civil Engineering Research Foundation. 1996. Level of Investment Study: U.S. Air Force Facilities and Infrastructure Maintenance and Repair. Washington, D.C.: Civil Engineering Research Foundation.

Collendar, Stanley E. 1995. The Guide to the Federal Budget, Fiscal Year 1996. Washington, D.C.: The Urban Institute Press.

Cotts, David F., and Michael Lee. 1992. The Facility Maintenance Handbook. Washington, D.C.: American Management Association.

Council on Environmental Quality and Office of Management and Budget. 1995. Improving Federal Facilities Cleanup. Report of the Federal Facilities Policy Group. Washington, D.C.: Government Printing Office.

DOT (U.S. Department of Transportation). 1996. Mechanical System Base-Line Testing, U.S. Department of Transportation, Nassif Building, Washington, D.C., 31 March 1996. Report by Summer Consultants, McLean, Virginia.

DOT. 1996. Mechanical System Base-Line Testing, U.S. Department of Transportation, Nassif Building, Washington, D.C., June 1996. Report by Summer Consultants, McLean, Virginia.

Earl, Richard W. 1997. The condition assessment survey: a case study for application to public sector organizations. Pp. 277–286 in Infrastructure Condition Assessment: Art, Science and Practice, Mitsuru Saito, ed. New York: American Society of Civil Engineers.

FCC (Federal Construction Council). 1988. Budgeting for Maintenance and Repair of Facilities. Technical Report No. 88. Consulting Committee on Operations and Maintenance. Washington, D.C.: National Academy Press.

FCC. 1991. Procedures Used to Assess the Condition of Federal Facilities. Technical Report No. 110. Consulting Committee on Operations and Maintenance. Washington, D.C.: National Academy Press.

FFC (Federal Facilities Council). 1996. Budgeting for Facilities Maintenance and Repair. Technical Report No. 131. Standing Committee on Operations and Maintenance. Washington, D.C.: National Academy Press.

FFC. 1997. Federal Facilities Beyond the 1990s: Ensuring Quality in an Era of Limited Resources. Technical Report No. 133. Standing Committees on Design and Construction and Organization and Administration. Washington, D.C.: National Academy Press.

GAO (General Accounting Office). 1990. NASA Maintenance: Stronger Commitment Needed to Curb Facility Deterioration. Report to the Chair, Subcommittee on VA, HUD and Independent Agencies, Committee on Appropriations, U.S. Senate. NSIAD-91-34. Washington, D.C.: Government Printing Office.

GAO. 1991. Federal Buildings: Actions Needed to Prevent Further Deterioration and Obsolescence. Report to the Chairman, Subcommittee on Public Works and Transportation, U.S. House of Representatives. GGD-91-57. Washington, D.C.: Government Printing Office.

GAO. 1993. Aging Federal Laboratories Need Repairs and Upgrades. Testimony. T-RCED-93-71. Washington, D.C.: Government Printing Office.

GAO. 1993. Budget Issues: Incorporating an Investment Component in the Federal Budget. AIMD-94-40. Washington, D.C.: Government Printing Office.

GAO. 1994. Financial Management: Army Real Property Accounting and Reporting Weaknesses Impede Management Decision-Making. Letter Report. AIMD-94-9. Washington, D.C.: Government Printing Office.

GAO. 1995. Budget Issues: The Role of Depreciation in Budgeting for Certain Federal Investments. AIMD-95-34. Washington, D.C.: Government Printing Office.

GAO. 1995. Depot Maintenance: Some Funds Intended for Maintenance Are Used for Other Purposes. Report to Congressional Committees. NSIAD-95-124. Washington, D.C.: Government Printing Office.

GAO. 1995. DoD Budget: Potential Reductions to Operation and Maintenance Programs. Briefing Report. NSIAD-95-200BR. Washington, D.C.: Government Printing Office.

GAO. 1995. National Parks: Difficult Choices Need to be Made About the Future of the Parks. Chapter Report. RCED-95-238. Washington, D.C.: Government Printing Office.

GAO. 1995. National Parks: Difficult Choices Need to be Made About the Future of the Parks. Testimony. T- RCED-95-124. Washington, D.C.: Government Printing Office.

GAO. 1996. Operation and Maintenance Funding: Trends in Army and Air Force Use of Funds for Combat Forces and Infrastructure. Report to the Chairman, Subcommittee on National Security, Committee on Appropriations, U.S. House of Representatives. NSIAD-96-141. Washington, D.C.: Government Printing Office.

GAO. 1996. Military Bases: Opportunities for Savings in Support Costs Are Being Missed. Letter Report. NSIAD-96-108. Washington, D.C.: Government Printing Office.

GAO. 1996. VA Health Care: Opportunities to Increase Efficiency and Reduce Resource Needs. Testimony. T-HEHS-96-99. Washington, D.C.: Government Printing Office.

GAO. 1996. Defense Infrastructure: Costs Projected to Increase Between 1997 and 2001. Letter Report. NSIAD-96-174. Washington, D.C.: Government Printing Office.

GAO. 1996. State Department: Millions of Dollars Could be Generated by Selling Unneeded Real Estate Overseas. Testimony. NSIAD-96-195. Washington, D.C.: Government Printing Office.

GAO. 1996. NASA Infrastructure: Challenges to Achieving Reductions and Efficiencies. Report to the Chair, Subcommittee on National Security, International Affairs, and Criminal Justice, Committee on Government Reform and Oversight, U.S. House of Representatives. NSIAD-96-187. Washington, D.C.: Government Printing Office.

GAO. 1997. Budgeting Issues: Budgeting for Federal Capital. Chapter Report. Report to the Chair, Committee on Government Reform and Oversight, U.S. House of Representatives. AIMD-97-5. Washington, D.C.: Government Printing Office.

GAO. 1997. Defense Infrastructure: Demolition of Unneeded Buildings Can Help Avoid Operating Costs. Report to the Chair, Subcommittee on Military Installations and Facilities, Committee on National Security, U.S. House of Representatives. NSIAD-97-125. Washington, D.C.: Government Printing Office.

GAO. 1997. General Services Administration: Downsizing and Federal Office Space. Testimony. T-GGD-97-94. Washington, D.C.: Government Printing Office.

GAO. 1997. Military Bases: Cost to Maintain Inactive Ammunition Plants and Closed Bases Could Be Reduced. Letter Report. NSIAD-97-56. Washington, D.C.: Government Printing Office.

GAO. 1997. High Risk Series: Defense Infrastructure. Letter Report. HR-97-7. Washington, D.C.: Government Printing Office.

GAO. 1998. Deferred Maintenance Reporting: Challenges to Implementation. Report to the Chairman, Committee on Appropriations, U.S. Senate. AIMD-98-42. Washington, D.C.: Government Printing Office.

Giorgione, Paul. 1994. Determining Navy Real Property Maintenance Requirements. Washington, D.C.: Department of the Navy.

Hawkins, Jeffrey, William Moore, and Trevor Neve. 1990. Managing Real Property Maintenance: Meeting the Challenge of Declining Budgets. McLean, Va.: Logistics Management Institute.

Institute for Water Resources. 1995. Living Within Constraints: An Emerging Vision for High Performance Public Works. Alexandria, Va.: U.S. Army Corps of Engineers.

Kaiser, Harvey H. 1989. The Facilities Manager Reference: Management, Planning, Audits, Estimating. Alexandria, Va: Association of Higher Education Facilities Officers.

Kaiser, Harvey H. 1993. The Facilities Audit: A Process for Improving Facilities Conditions. Alexandria, Va.: Association of Higher Education Facilities Officers.

Kaiser, Harvey H., and Jerry S. Davis. 1996. A Foundation to Uphold: A Study of Facilities Conditions at U.S. Colleges and Universities. Alexandria, Va.: Association of Higher Education Facilities Officers.

Lemer, Andrew. 1996. Infrastructure obsolescence and design service life, Journal of Infrastructure Systems 2(4): 153–161.

McGraw-Hill Encyclopedia of Science and Technology. 1997. "Nondestructive Testing," 8th ed., vol. 12, pp. 32–37. Chicago, Ill.: Lakeside Press.

Molof, A.H. and C. J. Turkstra, editors. 1984. Infrastructure: Maintenance and Repair of Public Works. New York: New York Academy of Sciences.

National Aeronautics and Space Administration. 1996. Reliability Centered Maintenance Guide for Facilities and Collateral Equipment. Washington, D.C.: NASA.

National Performance Review. 1993. From Red Tape to Results: Creating a Government That Works Better and Costs Less. Mission-Driven, Results-Oriented Budgeting. Washington, D.C.: Government Printing Office.

NRC (National Research Council). 1985. Building Diagnostics, A Conceptual Framework. Building Research Board, National Research Council. Washington, D.C.: National Academy Press.

NRC. 1988. Electronically Enhanced Office Buildings. Building Research Board, National Research Council. Washington, D.C.: National Academy Press.

NRC. 1990. Committing to the Cost of Ownership: Maintenance and Repair of Public Buildings. Building Research Board, National Research Council. Washington, D.C.: National Academy Press.

NRC. 1991. Pay Now or Pay Later: Controlling Cost of Ownership From Design Throughout the Service Life of Public Buildings. Building Research Board, National Research Council. Washington, D.C.: National Academy Press.

NRC. 1993. In Our Own Backyard: Principles for Effective Improvement of the Nation's Infrastructure. Building Research Board, National Research Council. Washington, D.C.: National Academy Press.

NRC. 1993. The Fourth Dimension in Building: Strategies for Minimizing Obsolescence. Building Research Board, National Research Council. Washington, D.C.: National Academy Press.

NRC. 1994. Aeronautical Facilities: Assessing the National Plan for Aeronautical Ground Test Facilities. Aeronautics and Space Engineering Board, National Research Council. Washington, D.C.: National Academy Press.

NRC. 1994. Space Facilities: Meeting the Future Needs for Research, Development and Operations. Aeronautics and Space Engineering Board, National Research Council. Washington, D.C.: National Academy Press.

NRC. 1995. Measuring and Improving Infrastructure Performance. Board on Infrastructure and the Constructed Environment, National Research Council. Washington, D.C.: National Academy Press.

Neely, Edgar S. and Robert Neathammer. 1991. Life-cycle maintenance costs by facility use, Journal of Construction Engineering and Management 117(2): 310–320.

NSTC (National Science and Technology Council). 1995. National Planning for Construction and Building R&D. NISTIR 5759. Committee on Civilian Industrial Technology, Subcommittee on Construction and Building. Washington, D.C.: Government Printing Office.

NSTC. 1995. Construction and Building: Federal Research and Development in Support of the U.S. Construction Industry. Committee on Civilian Industrial Technology, Subcommittee on Construction and Building. Washington, D.C.: Government Printing Office.

Occupational Safety and Health Administration. 1995. Indoor Air Quality Investigation, Nassif Building in Washington, D.C. Salt Lake City Technical Center Report.

O'Hara, Thomas E., James L. Kays, and John V. Farr. 1997. Installation status report, Journal of Infrastructure Systems 3(2): 87–92.

OMB (Office of Management and Budget). 1994. Government Performance and Results Act Implementation. Circular A-11. Washington, D.C.: Government Printing Office.

OMB. 1997. Analytical Perspectives, Budget of the United States Government, Fiscal Year 1998. Washington, D.C.: Government Printing Office.

Peters, Katherine M. 1997. Funding the fleet, Government Executive 29(1): 42–45.

Petze, John D. 1996. Investing in Facility Automation: Improving Comfort, Air Quality, Building Management and the Bottom Line. Manchester, N.H.: Teletrol Systems, Inc.

Rugless, J. Michael. 1993. Condition assessment surveys, Facilities Engineering Journal 21(3): 11–13.

Sanford, Kristen and Sue McNeil. 1997. Data modeling for improved condition assessment. Pp. 287–296 in Infrastructure Condition Assessment: Art, Science, and Practice, Mitsuru Saito, ed. New York: American Society of Civil Engineers.

Shen, Yung-Ching and Dimitri A. Grivas. 1996. Decision-support system for infrastructure preservation, Journal of Computing in Civil Engineering 10(1): 40–49.

Teicholz, Eric, and Takehiko Ikeda. 1995. Facility Management Technology Lessons from the U.S. and Japan. Norcross, Ga.: Engineering and Management Press.

Uddin, Waheed, and Fazil Najafi. 1997. Deterioration mechanisms and non-destructive evaluation for infrastructure life-cycle analysis. Pp. 524–533 in Infrastructure Condition Assessment: Art, Science, and Practice, Mitsuru Saito, ed. New York: American Society of Civil Engineers.

USACIR (U.S. Advisory Commission on Intergovernmental Relations). 1993. High Performance Public Works: A New Federal Infrastructure Investment Strategy for America. SR-16. Washington, D.C.: U.S. Advisory Commission on Intergovernmental Relations.

USACIR. 1996. The Potential for Outcome-Oriented Performance Management to Improve Intergovernmental Delivery of Public Works Programs. SR-21. Washington, D.C.: U.S. Advisory Commission on Intergovernmental Relations.

USACE (U.S. Army Corps of Engineers). 1990. Development of the BUILDER Engineered Management System for Building Maintenance: Initial Decision and Concept Report. TR-M-90/19. Construction Engineering Research Laboratories. Champaign, Ill.: Construction Engineering Research Laboratories.

USACE. 1995. Development of Condition Indexes for Building Exteriors. Report TR 95/30. Construction Engineering Research Laboratories. Champaign, Ill.: Construction Engineering Research Laboratories.

Urban Institute. 1994. Issues in Deferred Maintenance. Washington, D.C.: U.S. Army Corps of Engineers.

Uzarski, Donald R., Philip A. Weightman, Samuel L. Hunter, and Donald E. Brotherson. 1995. Development of Condition Indexes for Building Exteriors. Champaign, Ill.: U.S. Army Construction Engineering Research Laboratories.

Uzarski, Donald R. and Dana Finney. 1997 BUILDER-managing buildings, The Military Engineer 89(586): 36–37.

Uzarski, Donald R., and Lawrence A. Burley. 1997. Assessing building condition by the use of condition indexes. Pp. 365–374 in Infrastructure Condition Assessment: Art, Science, and Practice, Mitsuru Saito, ed. New York: American Society of Civil Engineers.

Wieman, Clark. 1996. Downsizing infrastructure. Technology Review 99: 49–55.